Forestry Commission Bulletin 63

Census of Woodlands and Trees 1979–82

G. M. L. Locke, B.Sc.
Census Officer, Forestry Commission

LONDON : HER MAJESTY'S STATIONERY OFFICE

ISBN 0 11 710202 4

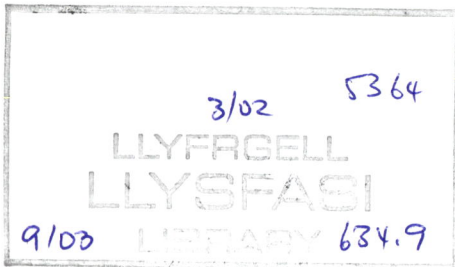

Contents

page

Preface   iv

Acknowledgements   v

PART I   PURPOSE AND METHODS

1   Previous Forestry Commission Woodland and Tree Surveys   2

2   Background to the 1979–82 Survey   5

3   Availability of Existing Information   7

4   Work Preparatory to the Start of the Census   11

5   Sample Selection – Woodlands   17

6   Sample Selection – Non-Woodland Trees   20

7   Air and Ground Assessment   24

8   Data Transmission and Storage   31

9   Calculation and Presentation of the Results – Woodlands   35

10   Calculation and Presentation of the Results – Non-Woodland Trees   40

PART II   RESULTS

11   Total Woodland Area   44

12   Private Woodland Area   56

13   Forestry Commission Woodland Area   66

14   Non-Woodland Trees   74

15   Standing Volume   80

16   Volume Increment   85

PART III   COMPARISONS WITH PAST SURVEYS

17   Methods and Results – Woodlands   90

18   Methods and Results – Non-Woodland Trees   97

PART IV   CONCLUSIONS

19   Conclusions   102

APPENDICES   103

Preface

This Report has been compiled by G M L Locke, the present Census Officer of the Forestry Commission. It describes the survey methods adopted for the Census of Woodlands and Trees carried out between 1979 and 1982, discusses the main results of the investigation and compares them with those of past surveys. It supplements the area and volume results which have already been published for counties and Conservancies in England and Wales, for Regions and Conservancies in Scotland, for the three countries and for Great Britain as a whole. In this Report considerable emphasis has been laid on the background to the Census and on the survey methods which finally evolved from the original tentative thoughts on design. Many of the systems finally adopted were new to forest inventory methods in this country; as such they were not introduced without problems and it has been considered advisable to highlight some of these difficulties for the benefit of others who might wish to consider similar lines of enquiry.

The survey was carried out by the Census Branch which at the time formed part of the Field Survey Branch of the Research and Development Division of the Forestry Commission. Mr K P Thallon, the then Chief Field Survey Officer, had overall responsibility for the Census, was involved in much of the early planning work and gave detailed guidance throughout. Mr W T Waters took up the post of Census Officer in November 1979 and was responsible for the day-to-day running of the operation. Mr A C Miller and Mr R W Twallin were the two Head Foresters in charge of the field teams with the latter also responsible for many of the office procedures.

ACKNOWLEDGEMENTS

Thanks are due to the many landowners and their agents in permitting entry of the Census Surveyors on to their land and also for providing information on their woodlands and trees; also to the Director General of the Ordnance Survey for the help and advice given by his staff and for permission to publish certain maps and other material contained in this Report.

In a survey of this size and complexity it is impossible to give credit by name to the many members of staff of the Forestry Commission who helped at the various stages of the field and office work. However, their contribution, and particularly that of the surveyors, if not individually acknowledged is nevertheless fully and gratefully recognised. The author, however, is particularly indebted to the following for their help on the compilation of this Report:

Mr A J Grayson, the current Director of Research for the Forestry Commission, Mr W T Waters, Mr A C Miller, Mr J C Proudfoot, Mr J M Christie, Mr R Gill, who largely contributed Chapters 8, 9 and 10 and Mr K Rennolls for his help with Chapters 5 and 6.

All gave freely of their time and knowledge in providing background information and helpful criticism of the Report at its draft stage. Nonetheless, the final responsibility for this Report remains with the author.

Part I

PURPOSE AND METHODS

CHAPTER 1

Previous Forestry Commission woodland and tree surveys

The collection of statistics on the country's stock of woodlands and trees is a responsibility which is laid upon the Forestry Commission by the Forestry Act of 1919, the Act which brought the Forestry Commission into being. For this reason the Forestry Commission, as one of its duties of data collection, carries out assessments at intervals of the national position. The 1919 Act does not specify how the data collection should be undertaken, nor how intensive it should be; these are decisions which the Forestry Commission usually has to make itself, although normally in consultation with other interested bodies.

There have been a number of national surveys since 1919 with the interval between them normally being 15–20 years. In each the objectives have differed slightly but significantly and in consequence, the nature of the information that has been collected, and the precision of the results, have also differed.

1924 Census

The first Census was in 1924 which, because of manpower shortages, was based on questionnaires sent to owners. The purpose of the Census was to establish the then current position after the very extensive fellings of the First World War. It suffered from the defects inherent in such surveys namely, variation in the fullness and accuracy of the returns and local differences in interpretation. It did, however, indicate the extent of the woodland devastation that had taken place, and consequently the amount of rehabilitation required, and to this end it fulfilled its prime function. The minimum wood size chosen for this survey was two acres (0.8 of a hectare).

1938 Census

The next Survey was that undertaken in 1938 when the possibility of war was causing increasing concern as to whether there were adequate stocks of home grown timber in Britain to meet the emergency, bearing in mind the very high proportion of our needs that were already coming from abroad. The

intention of this survey was to undertake a census of woods of five acres (2.0 hectares) and over throughout the country with the aim of being able to locate crops of specific species, age and sizes with the minimum of difficulty. Unfortunately, because of the amount of detail that was being collected, and consequently the relatively slow progress, the survey was only partially completed by the outbreak of war. Those parts of the country which had not been assessed were dealt with by means of a sampling survey which, of course, provided the necessary area and volume estimates but not the locational information. The choice of five acres as the minimum size of wood is interesting and was the outcome of trial surveys to test methods. Because of the scattered nature of woodland it was found that there is a minimum size below which the gains from having increased information are outweighed by the cost of obtaining it. For timber production purposes interest was also obviously bound to be centred on the medium and larger sized blocks.

1947 Census

The next Census to be carried out was the 1947–49 Census which is probably the most complete woodland survey ever to have been undertaken in this country. Its purpose, like that of the one in 1924, was to establish how much productive woodland remained, and also the extent of the devastation that had been brought about by wartime felling. Again, however, we have to remember the conditions that reigned at that time. It was a period of austerity with imports restricted, felling was still continuing at a substantial level and closely controlled licensing of cutting was still in force.

It therefore seemed sensible to adopt one of the main principles of the 1938 Census, that there should be a record of the contents of every woodland block to provide the desired locational information, but to restrict the amount of data collected in each so as to speed up the process of collection. The classification adopted was, however, based on the types of crops that were likely to be encountered and so included forest types such as Devastated, and the subdivision of Felled into crops that were cleared before or after the outbreak of war, to cover woodlands that had been affected by felling to the degree whereby they no longer could be allocated to one of the more conventional categories such as High Forest, Coppice or Scrub.

The minimum size of wood chosen was five acres, for the same basic reasons as in 1938, but it was appreciated at the outset of the survey that trees in woods under this minimum area, and also isolated hedgerow and park trees, would need to be dealt with at a later stage. The main survey was completed by 1949 and provided a detailed record of the position at that time. The fact that it was done in such detail makes it a modern Domesday Book for woodland, and so acts as a yardstick against which subsequent changes can be measured.

The processing of the data, however, created difficulties not only on account of the sheer volume of information but also the lack of sophisticated aids as we now know them. The details were recorded on forms by putting crosses in boxes, the information transferred to Hollerith punch cards and the data then listed and tabulated. This resulted, of course, in crops being identified only by their major characteristics so that, for example, the whole area of a crop of pine and spruce in 50:50 mixture was credited to the first species listed, and nothing to the second. It is probable that over a large number of crops this situation evened itself out but there is no certainty of this; a species such as elm, which is often present in woodland but seldom dominant, would certainly have its area under-represented.

1951 Survey

The associated survey of woods of between one and five acres, and of hedgerow and park trees, was carried out in 1951 in order to complete the picture. In both cases a sampling survey of fairly low intensity was adopted. In the case of woods of between one and five acres some work had been done during the main Census to determine the likely area in this category. A sample of 6 inch to one mile Ordnance Survey maps was examined and all woods of between one and five acres on them were measured to establish an area total for each country. Subsequently the field work was confined to assessing a sample of woods for the main crop characteristics such as forest type, species and age. In retrospect one can see reasons for the overall area being in doubt although its subdivision by forest type, etc., is probably reasonable enough. The occasion for doubt about the survey's reliability lies in the fact that the 6 inch maps available just after the war were of a variety of editions and in some cases the date of survey by the Ordnance Survey was inter-war while in

others it could date from well before the First World War. The map base was therefore out of date and possibly inaccurate. However, the field sample, although small, did not raise any doubts about the accuracy of the woodland area totals already established.

The hedgerow sampling used methods first adopted in 1938. The procedure to determine the volume of isolated trees and trees in woods of under one acre involved the assessment of all trees in these categories within sampling strips which were one mile long and two chains wide, an area of 16 acres (6.47 hectares). The intensity of sampling adopted was designed to produce a volume figure for Great Britain as a whole correct to ±10%. This precision was, in fact, attained for Great Britain and also for England, whereas the precision of the estimates for Wales and Scotland lay between 20 and 25 per cent. These were considered acceptable in view of the relatively small volumes involved in these two countries. The great bulk of the volume was known to occur in England and, in particular in central and southern England. The intensity of sampling required to produce these estimates was about 1:8000, that is one hectare for every 8000 hectares of land and inland water area. This is a fairly small sample but it provided an overall estimate precise enough for the purpose for which it was intended and, in addition, supplied reasonably accurate information on species and size class distributions within individual countries. It also indicated the relationship between the volume in this category and that in small woods and woods of five acres and over.

1965 Census

The survey in 1965 was needed because the period since 1947 had been one of tremendous change. Much had been done to try and repair the ravages of the wartime and post-war fellings by restocking and, in addition, a very substantial area had been afforested. Some of the changes that had taken place were known in detail, others could only be estimated, and it was realised that the time had come to make a new assessment. The concern was primarily about the accuracy of the production forecasts being prepared at that time because many wood-using industries utilise timber of a particular species or group of species, or material cut to particular specifications. If new industries were to be set up or helped to expand, it was essential to have information on the potential yield of wood within the catchment areas. The 1965 Census was therefore undertaken primarily to supply information for marketing regions, usually groups of counties, and consequently differed from previous surveys in that the production of estimates by county, whether it be area or volume, was no longer a major consideration. This was because the major wood using industries have catchment areas which are usually based on haulage distance of the order of 50-100 miles and consequently are independent of local authority boundaries.

There were two other departures from previous practice. The first was that the minimum wood size was reduced to one acre (0.4 of a hectare); the planting grant regulations at the time permitted payment down to this size and the limit still accorded with the minimum adopted in the 1951 Survey. The second change was that the Forestry Commission by this time had a record of its own crops contained in a database, which obviated the need for a survey of State woodlands.

The survey method utilised was one which had been adopted for a broadleaved pulpwood survey which had been carried out in 1963. The procedure involved the inspection of all private woodlands contained within selected kilometre grid squares distributed throughout the country. The one inch to one mile Ordnance Survey maps were used to identify the sample squares which contained woodland. In the case of those squares that were apparently lacking any woodland over one acre in extent, local Forestry Commission staff were asked to confirm that the map and ground situation were the same. The use of kilometre grid squares had the added advantage that estimates of the woodland position, and consequently likely volume production, could be made for any defined area and was thus independent of local authority or any other boundary limitations.

The results of the survey, although worked up on a county basis, were aggregated to produce regional information and, from there, summarised to national level. The aim of the survey was to provide figures of private woodland area in each region to a precision of $\pm 2\frac{1}{2}\%$ bearing in mind that subdivision of these figures by forest type was likely to result in a precision between $\pm 5\%$ and $\pm 7\frac{1}{2}\%$, and further subdivision by species, age class, etc., would result in

their precision being no better than $\pm 10\%$ for each subset. It was intended, however, that the private woodland area for the country as a whole would be assessed to within $\pm 1\%$. These requirements were generally met. Nevertheless, the level of sampling was not particularly high and the results were insufficiently accurate to be used as estimates at county level; in any case, Forestry Commission records were not maintained on a county basis on the computer at that time.

Although every sample kilometre grid square that was known to have, or deemed likely to have, woodland on it was visited and assessed, there were, of course, many squares that were thought to be blank, and so not visited, making it possible that some woodland was missed. There was, at that time, only limited up-to-date cover of aerial photography so that this possible source of verification was not available.

A hedgerow survey was carried out contemporaneously with the main survey using a sampling method and intensity similar to that used in the 1951 Survey. One third of the samples measured in 1951 were re-selected for measurement and the remainder randomly chosen. By that time the majority of Ordnance Survey maps were based on the National Grid so the alignment of the new strips was changed from true North to Grid North. The major difference between the two surveys was, however, the fact that the area sampled was confined to England south of a line drawn between the Mersey and the Humber. This was for purely financial reasons since it was known that 65% of the total volume could be sampled whilst measuring only 40% of the number of strips assessed in 1951.

The subsequent 15 years to 1980 saw a major increase in woodland area as a result of the substantial planting programme of both the Forestry Commission and private owners, together with the arrival in this country of Dutch elm disease with its consequent spread and the devastation which attended it. The combination of major change and the lapse of years since the last census made it advisable that a new survey be carried out.

CHAPTER 2

Background to the 1979–82 survey

The decision to undertake a new Census of Woodlands and Trees is only taken after lengthy deliberation and there are usually a number of influencing factors. On this occasion none were individually of over-riding importance but in combination they determined not only the timing, but also the nature and the intensity of the assessment that was needed. The following aspects undoubtedly had a major bearing on the decision to undertake a new national woodland and tree inventory.

First, immediately after any Census there is a full and clear picture of the size of the woodland resource and its major features. Normally every year there is a certain amount of information available on woodland change such as estimates of the area of new planting and restocking, and current levels of volume removal from both thinning and felling, and it is therefore possible for quite a number of years to use such records to adjust the Census totals annually. This enables both known and expected change to be built into the totals with reasonable confidence for areas as large as countries. After a period, however, errors or omissions in the assumptions begin to compound themselves and each subsequent year sees a growing uncertainty about the accuracy of the adjusted data. The updating of information over time therefore becomes progressively more difficult at national level, and even more so where smaller land areas are concerned.

Secondly, there was obviously a growing awareness by the general public of changes that were taking place in the countryside. People had more leisure, travelled more and were taking an increasing interest in their environment. The spread of Dutch elm disease was an obvious cause for concern and by the early 1970s it had resulted in widespread death of elm in southern and central England. The passage of the disease was, in many ways, made more obvious as a result of elm being a tree of hedgerows rather than woodland and many of the tree planting schemes initiated around that time were prompted by the stark evidence of dead elm in the countryside. There was also a more general desire to help beautify the environment or hide specific industrial and other eyesores with trees.

Another subject of public concern was the changing agricultural landscape. Hedgerows were being grubbed up, copses were being felled and converted to agricultural land, and moorland was being reclaimed. The 1960s and 1970s were also a period when extensive road widening and motorway construction were taking place which resulted in some woods being much reduced in size or disappearing altogether. In contrast, the fact that a lot of new woodland has been planted as a consequence of motorway construction does not always register immediately and it takes some years before such plantings take on a woodland character. Housing development has accounted for a substantial loss of woodland, and will continue to do so, although remnants may be retained for amenity.

Furthermore both World Wars caused the removal of much of the good quality broadleaved timber leaving in many cases stands of poorer merchantable potential; most of these crops were also mature or over-mature. With the passage of time these crops were cleared, or were exploited, and hardwood timber merchants found it increasingly difficult to obtain supplies. The problem was probably not so much a lack of suitable material but rather one of growers not being able to sustain supplies at a stable level, particularly where specialist markets were involved. This difficulty is closely related to the inability of private owners to market efficiently because individual owners can seldom supply in quantity, resulting in the offer of parcels with a wide variation in quality.

If there were difficulties in marketing suitable material on the grower's side it was also true that the hardwood timber trade was at a crossroads. There was a changing pattern from the father and son business working with old machinery to a more modern, cost-conscious, profit orientated, business where the availability of future supplies was just as important as establishing future markets. There was therefore a need for capital investment to modernise sawmills and harvesting equipment. In 1977 it was estimated that such modernisation could cost up to a quarter of a million pounds for a small unit and over a million pounds for a large unit*. These factors together with a projected world shortage of hardwood supplies by the year 2000 obviously resulted in merchants seeking some assurance that the broadleaved supplies were not only adequate to meet current needs but also that the nature of the broadleaved growing stock would permit a level of sustained yield for a long enough period to justify the high costs of mill modernisation and rationalisation.

The local government reorganisation which took place in England and Wales in 1974, and in Scotland in 1975, resulted in the new authorities reassessing their positions in respect of land use planning in their areas and they naturally wanted to have statistics to enable them to fulfil this role. So far as trees and woodlands were concerned their requirements in the main were to be in a position to present their Planning Committees with a comprehensive picture of the woodlands in their area so that in the event of applications for felling licences being referred to them, or in the case of the imposition of Tree Preservation Orders, the officials could evaluate the particular woodland's contribution to amenity or conservation within the overall picture for a county or a district. Also, the granting of felling licences and the imposition of replanting conditions could be seen not as a series of one-off events but as part of a long term strategy to maintain tree cover or retain the characteristics of conservation areas. To meet these needs a number of local authorities had carried out, or were intending to carry out, their own woodland surveys which would have certainly served the aims and objects of the county in question but it was most unlikely that an adjoining county carrying out a similar type of survey would adopt the same criteria. The results of such surveys would therefore be incompatible and of no value in the compilation of regional or national statistics. It was equally obvious, however, that even if only the minimum requirements of local authorities were to be met it would mean that any survey carried out by the Forestry Commission would need to be much more intensive than that of 1965 which was aimed primarily at producing figures for groups of counties forming marketing regions.

Concern about the rate of change in the countryside and the apparent reduction of broadleaved trees in the landscape was also being voiced by a number of national bodies such as the Department of the Environment, the Countryside Commissions and the Tree Council. Bodies like the Countryside Review Committee set up by the Government in 1974 had asked questions of the Forestry Commission for which statistics were not available. As one of the functions of the Countryside Review Committee was to monitor environmental information about the countryside relating to change the suggestion that a national tree census be carried out had been mooted by their members. The Scottish Standing Committee on Rural Land Use had also noted the lack of up-to-date statistics on certain forestry aspects and it was clear that the need for a new assessment was pressing.

---

* In 1985 terms these figures represent sums of over half a million and two and a quarter million pounds respectively.

CHAPTER 3

Availability of existing information

Arising from the demand for more comprehensive and up-to-date statistics consideration was given in 1976 to the design of a new survey that would meet the Forestry Commission's own need for data and would also incorporate as many of the requirements of other users as were practicable.

The first task was to consider what information was already available about the woodlands and trees in Great Britain and whether it could be utilised directly or indirectly as part of the national census. Two major sources were known to exist, the Forestry Commission's own database, and information relating to the Dedication and Approved Woodland Schemes.

The Forestry Commission's own computerised database had been gradually built up over a period of years and was derived from surveys carried out in its own woodlands. To satisfy the requirements of forecasting wood production the information needs not only to be up-to-date, and this is done by annual amendment of the records for change, but also fairly detailed in respect of crop type, species proportions, age, yield class, etc. The information was converted from imperial units to metric in 1970/71 and the opportunity taken at that time to ensure that the basic information was complete and as accurate as possible. Forest survey work is a continuing process in the Forestry Commission, as is the annual amending process, and it was known that the information relating to this ownership category would be more than adequate to meet the needs of the Census without any further work being done. Forestry Commission woods at the time accounted for about 40 per cent of the total woodland area of Great Britain but its share in numbers of trees and volume in trees outside the forest was negligible.

The second source of data was the Plan of Operations relating to each individual Dedication or Approved Woodland Scheme. Owners in receipt of grant-aid under these Schemes were obliged to prepare a Plan of Operations which contained information about the crops which were covered by the Scheme and included detailed proposals for management for the first five-year period and less detailed proposals for the next five. Every five years

the Plan was reviewed and updated. Use had not been made of this information at the time of the 1965 Census for two reasons. The first was that at that time less than a third of the total private woodland area was being worked under these Schemes and secondly, and more importantly, the method of survey adopted in 1965 whereby whole kilometre squares were assessed, would have meant that accessing the information in the files would have been relatively time consuming and would still not have precluded the necessity for a ground visit to collect information on crop height and diameter. During the late 1960s and early 1970s there was a major increase in the area included in the Dedication and Approved Woodland Schemes and by 1975 the proportion of private woodland included under Plans of Operations was rising fairly rapidly and was obviously a major source of valuable information. Up until this stage no attempt had been made to bring this private woodland information from the individual Scheme files in which it was held at Forest District or Conservancy level on to the computer and it was fully realised that this would constitute a major operation. However, the necessity for such a data transfer had already been under consideration by the Secretariat Division of the Forestry Commission so that it would be in a better position to answer queries about the progress of the Schemes posed by Members of Parliament, other Government bodies, local authorities, etc. The Harvesting and Marketing Division had also expressed an interest in that it could help to improve the accuracy of the private woodland volume forecast.

From the Census point of view the information available suffered from certain limitations. The first was that the amount of detail in which private owners had been asked to collect their data was much less, and the form of presentation much simpler, than that used by the Forestry Commission for its own woodlands. Consequently it was known that the data would not be complete and would also vary in accuracy. The second aspect was that the data for individual Schemes are only updated at five yearly intervals and sometimes longer if little or no change has taken place in the previous five year period.

Thirdly, certain desirable data such as the sub-division of the plantable land category including scrub and felled, were not recorded. Nevertheless it was felt that the statistics contained in the Plans of Operations would be sufficiently accurate to enable these estates to be excluded from the survey side of a future national inventory and the decision was made to examine practical methods of computerising the private woodland records. Although it was decided in 1976 to proceed with the transfer of detail from the files to computer, manpower restrictions made it obvious that this was not a task that could be completed quickly; however, arrangements were made for the work to be put in hand and the results were to be available in time for them to be incorporated in any future Census. These private woodland records cover only woodlands and do not contain any information on numbers or volume of trees outside woodland.

There had already been a series of informal discussions between members of the Research and Development Division of the Forestry Commission and those of environmental agencies and other organisations in order to gain knowledge of their data requirements. To formalise the situation a Steering Committee was set up in June 1976 with the objectives of:

a. ensuring that the definitions adopted and data collected would be compatible with those already existing for Forestry Commission and Dedicated and Approved woodlands;

b. to find out in more detail the uses which organisations other than the Forestry Commission would make of the data, to determine the size of the region (county or district) for which data would be required, and the desired accuracy of the results;

c. to give an estimate of the likely cost of a full national census designed to provide the required data.

The Steering Committee included representatives of the Department of the Environment, the Countryside Commission and Nature Conservancy Council with the aim of ensuring that the interests of these and other organisations were taken into account at the planning stages of the Census, and also to provide advice on certain aspects where these organisations had particular knowledge. Other bodies, such as the Countryside Commission for Scotland, although not represented on the Committee were kept advised of progress.

A series of field trials were also undertaken to test various aspects of survey design and to ensure that where it was proposed that additional information should be collected this could be done both practically and cost effectively under field conditions. Most of this trial work was done in West Sussex for four major reasons:

a. it is a heavily wooded county, contains virtually all the forest types that were likely to be recognised in a Census and has a wide range of species and ages;

b. the Committee had the full co-operation of West Sussex County Council on access to information and other practical issues;

c. there was recent aerial photography of the whole County at 1:12 000 scale permitting the existing map coverage to be checked and updated;

d. the County lies close to the Alice Holt Research Station of the Forestry Commission from which the trial work was carried out.

Much of this development work was undertaken by geography students of Lanchester Polytechnic under the direction of a Forestry Commission Head Forester. Each student spent about 12 months with the Forestry Commission and used the work they were doing as the basis of a project for their degree.

Certain basic facts and conclusions emerged both from the development work and the discussions of the Steering Committee. It became clear that a complete survey of the woodland resource could not be countenanced. The cost would have been prohibitive and unless the whole task could have been completed reasonably quickly would be of doubtful value owing to the continuing process of fairly rapid change. This then meant a sampling survey and, because of the high costs of ground assessment, there would need to be a greater dependence on aerial photography than in the past. It was also apparent that because of the increased interest in trees in general as distinct from woodland much more work would need to be done on determining the numbers and sizes of isolated trees and trees in clumps or avenues as well as collecting information on their actual location within the environment. The survey methods required to assess woodlands and non-woodland trees would undoubtedly differ radically but few practical problems were foreseen to running the two assessments in parallel.

On the woodland side the first and major consideration was how best to obtain a fairly precise estimate of total woodland area. With the Forestry Commission woodlands and Dedicated and Approved woodlands already catered for the problem was to determine the area of 'Other' private woodland. Interest naturally concentrated on the 1: 50 000 Ordnance Survey map series which had been issued as a new series in 1974.

It was appreciated that in most cases the new map edition had been produced from photographic enlargements of the one inch to one mile series which it replaced, with only some of the detail having been revised, but nevertheless the new map series was the most up-to-date national representation of wood-land available. The minimum woodland area considered for the Census at this juncture was one hectare with any woods smaller than this being included in the non-woodland category. Various methods of measuring the areas of woodland blocks at 1:50 000 scale were then tried using planimeters, dot grids or intercept methods i.e. measuring the length of woodland on a series of grid co-ordinate lines and then converting the percentage determined for the county into an area. It was also considered that it would be more efficient to assess both woodland and non-woodland features on the basis of stratified random samples based on drift geology and climatic factors which would represent zones of broadly uniform landscape and so be likely to have similar ecological and geographical characteristics.

Certain aspects were thought worthy of being considered in much greater detail than had been done in the past, for example the distinction between woodlands in private possession and those on common land because of the differing legislation involved. Where appropriate, consideration should also be given to recognising the occurrence of woodland in National Parks, Areas of Outstanding Natural Beauty, National and Local Nature Reserves, Sites of Special Scientific Interest, etc. The actual woodland features to be assessed on the ground would be the conventional ones of forest type, species proportion, stocking density and age, together with sufficient information on height, diameter, basal area, to enable volume to be determined.

On the non-woodland tree survey it was considered desirable to record the location of trees, i.e. urban, rural, isolated, in clumps or in belts, on boundaries or open grown; the type of boundary on which they occurred, e.g. hedgerow, wall, bank, etc., and whether the boundaries were roadside, waterside, field, garden, etc. Trees were to be recorded in terms of their species, age group, height, crown diameter and breast height diameter. Hedges, copses and belts were to be classified in terms of width and length. Health was also to be introduced as a characteristic.

These features constituted a formidable amount of data to be collected but it seemed at the time that they catered for the main interests of the members of the Steering Committee. Much of the desired information could of course be obtained directly from aerial photography, with the larger scale photographs providing more detail than the small scale. Nevertheless a considerable amount of ground survey would be necessary to establish the relationship between the data derived from air photographs and those derived from ground measurement, and also to collect information which the air photographs could not reveal.

In May 1977 a small Working Group was set up within the Forestry Commission with the following terms of reference:

a. to ascertain the minimum information and the levels of precision required from a Census of Woodlands by the Forestry Commission, the Home Timber Merchants' Associations, Department of the Environment, Nature Conservancy Council, Countryside Commission, local authorities, and any other appropriate bodies;

b. to determine the most cost effective way in which a Census Revision could be undertaken to provide the information required under a. and how such a Census should be undertaken and updated on an ongoing basis;

c. to determine the cost of such a Census and who may be expected to contribute to the cost and to what extent;

d. to recommend the extent of the Forestry Commission's investment in such a Census.

This Working Group held several meetings at which it considered the recommendations of the Steering Committee, discussed with a variety of organisations the nature of the data they required and produced a draft County Report so that the tabular statements could be finalised. In April 1978 the decision to undertake a new Census was taken by the Heads of Divisions and ratified by the Forestry Commissioners in May 1978. The following procedures were to constitute the main framework of the Census.

First, it was to be based on a system of sampling and was to include all trees in Great Britain except those in Forestry Commission forests and Dedicated and Approved woodlands, for all of which data were already available, and for certain island areas where the tree density was very low. Trees in towns that were not readily accessible would, also, be excluded.

Second, the survey would be divided into two main parts:

a. woodland (0.25 of a hectare and over);

b. hedgerows, park trees and clumps (under 0.25 of a hectare);
    i.   rural areas;
    ii. urban areas.

Where appropriate the data to be recorded would include:

Woodland	Non-woodland trees
Area	Species
Forest type	Volume
Species	Location
Age	Health and life expectancy
Volume	

Third, it was considered that the basic sampling unit should be the County in England and Wales and the District in Scotland and that the precision of the area estimate should be ±10%. This level of precision would provide accurate basic data for an area the size of a Conservancy and the detailed information, although less precise, would generally still have an acceptable level of accuracy. When the data from the Forestry Commission forests and Dedicated and Approved woodlands were added to the Census data the precision of the woodland statistics would be considerably higher than that yielded by the Census alone.

Fourth, the data would be summarised by Counties in England and Wales, by major Districts and by Regions in Scotland, Conservancies, Countries and for Great Britain as a whole, and possibly also for special areas such as National Parks.

The only major departure from the proposals of the Steering Committee lay in the adoption of 0.25 of a hectare as the minimum size of woodland to be considered in place of the limit of 1.0 hectare previously considered. The lower figure was selected because it was at that time the minimum size of block eligible for grant aid under the Small Woodland Scheme.

The way was now clear to start the more detailed planning for the new Census.

CHAPTER 4

Work preparatory to the start of the census

Although the decision to undertake a new Census of Woodlands had been taken by May 1978 there still remained a number of investigations which needed to be completed before field work could begin. These can be considered under three headings:

a. how was the area of 'Other' private woodland in each county or district to be determined?

b. how were the air photographs and ground survey approaches to be integrated so as to obtain the maximum benefit from both systems?

c. how was the detail collected at the time of the ground survey to be recorded and what programming would be required to produce the necessary tabular summaries?

It will be evident that considerable attention had already been directed to the use of the 1:50 000 map as a means of establishing woodland area and as field trials progressed certain aspects became clearer.

The main advantage of using this map series was that at the time it formed the most up-to-date presentation of woodland available on a nationwide basis; woodland was certainly shown down to an area of one hectare and often down to 0.25 of a hectare and below if conspicuous. The map series was also available on astrafoil – a stable base medium – which overcame the problem of using paper prints for area measurement as they are inevitably subject to distortion as a result of differential paper shrinkage.

There were, however, also a number of disadvantages related to its use. The first concerned the accuracy of representation of woodland blocks on the 1:50 000 maps. This was generally very good for the larger blocks and comparison of the areas on these maps with the corresponding area on the ground, subsequently measured at 1:10 000 or 1:10 560 scale, gave similar results. Correlation was poorer with blocks of less than 20 hectares due to a number of factors which included the difficulties of measuring such blocks accurately at 1:50 000 scale, the fact that woodland boundaries on these smaller blocks are sometimes indeterminate as they are not always bounded by sharp features and also that some map detail such as roads are sometimes shown

on the map in diagrammatic rather than plan form, and may be exaggerated in width; consequently woods which border roads may be displaced sideways or, in the case of larger woods, actually reduced in size and consequently in area. Alternatively, some woodland blocks may appear on the map which on inspection prove to be lines of trees or narrow belts. Any measurement errors arising from these factors have a proportionately greater effect on a small block than on a large one and it was appreciated that a careful check would need to be made on the various reductions and enlargements, especially in the smaller sized blocks, to ensure that appropriate adjustment could be made to the area total. Furthermore, there were very few map sheets that had been surveyed recently and the remainder inevitably were out of date to some degree. Also, revision by the Ordnance Survey does not necessarily take place over a whole sheet at one time and even the most recently published editions are never completely up-to-date either for this reason or because of the lapse of time between resurvey and publication. However, having weighed up the various advantages and disadvantages of using the 1:50 000 map series, discussed the subject with members of the Ordnance Survey and looked at other possible alternatives, the conclusion reached was that the use of this series was likely to be the most effective solution to the problem of establishing woodland area.

The next task was to establish the best way of recording the occurrence of all the woodland blocks on a 1:50 000 sheet and also determining the area of each one by measurement. Most of the trial work to date had used planimeters or dot grids for area measurement. However, the use of these methods was a sizeable task for even one map sheet, so to consider doing a national survey by this method with nearly 200 sheets involved was obviously not a practical solution. However, by this time the possibility existed of extracting the desired information using digital methods. Essentially these consisted of placing a map on a digitising table, reconciling the National Grid co-ordinates of the map corner and those of the digitising table and then tracing the boundaries of each woodland block on the map using a cursor. The grid co-ordinates of selected points on the boundary of the wood were noted or alternatively co-ordinates were taken automatically at specified time intervals. Each individual block was given a serial number and a datum point grid reference and the area correct to 0.1 of a hectare subsequently calculated. In order to ensure that no blocks were missed, and that there were no obvious errors, the results were produced on transparent plastic film which could then be overlaid on the original. This approach undoubtedly speeded up the process substantially and removed the need for the tedious measurement systems previously adopted.

Use of a digitising table, despite being more speedy and accurate, still had obvious difficulties when related to the sheer size of the task. For example it was estimated that the total number of woods to be measured in Great Britain was unlikely to be less than 100 000 and could easily be more than double that figure. An automatic method of scanning the blocks was therefore sought and a solution found in the use of the FASTRAK digitisers owned and operated by Laser-Scan Laboratories Ltd. of Cambridge. Programs which had been designed for digitising contours were found to be capable of measuring woodland areas after suitable adjustment. The FASTRAK system employs a very fine laser beam steered at high speed and is used to interrogate a film which in this instance was a reduced photographic negative of part of the original map. The co-ordinates of each cornerpoint of the map were first recorded as this removed any dependence on the precision of the photographic reduction factor or on the alignment of the negative in the holder. In this case the map was supplied to Laser-Scan in the form of a woodland plate, i.e. the only details shown on the astrafoil were the woodland blocks, the ownership notation and the grid co-ordinates of the sheet corners. The operator set the beam on the edge of a wood, and given the appropriate instruction the beam traced the outline of that wood and recorded the details of the boundary changes in terms of National Grid co-ordinates. As each wood was completed a second laser 'filled in' the appropriate area and removed it from the view of the scanner and the operator. Thus there is not only verification of the details but little possibility of repetition or omission. Where woodland boundaries were complicated the machine could be programmed to stop if there were choices, the imagery enlarged on a screen, and the machine operated manually until the difficult section had been completed. In this way information was built up for each County or, on occasion, District with each wood having its own individual serial number and a centroid grid reference. These details, together with the area of each

block, which had been calculated on an associated computer, were then transferred to the Forestry Commission.

Differing annotations on the astrafoil were used to distinguish Forestry Commission (F), Dedicated (D) and Approved (A) woodlands; 'Other' private woods were left unmarked. In England only the 'Other' private woodlands were digitised but by the time Wales and Scotland were ready to be dealt with it was decided that all three classes should be digitised so as to provide the full record. In addition therefore, for Wales and Scotland, any known woodland extension resulting from planting having taken place after the date of map revision was recorded together with any extensive unplanted or unplantable areas within Forestry Commission boundaries. In order to distinguish ownership classes which adjoined one another within a woodland block a line often had to be scribed on the astrafoil to mark the separation. The addition of this type of information did complicate the recording and the reading of detail and the use of reverse reading positive in the machine rather than a negative proved helpful on some occasions. A record on clear film was also produced which indicated the outline of each block and its ownership denoted by means of a specific line style. If more than one ownership class were digitised, differing line styles were used to distinguish them. This check plate could then be overlaid on the original map and the area of any individual block quickly determined from the listing provided.

In order to provide the system with the requisite data all Forestry Commission and Dedicated and Approved woodlands had first to be plotted on to paper copies of the maps and, after checking, the detail transferred to the astrafoil plates. This constituted a major task as the bulk of the record maps relating to Forestry Commission woodlands were still either at one inch to one mile or 1 : 10 000 scale and the mapped information for Dedicated and Approved woodlands usually at 1 : 10 000 scale and contained within individual files. Only occasionally had the information been transferred to 1 : 25 000 or other scale record maps. The task of transferring the required information to produce the master copies was therefore very substantial indeed and had to be undertaken as a special exercise for the Census at each of the Conservancy Drawing Offices with the work being spread over a period of about 18 months.

The ability to determine the total area of woodland from the maps, and to allocate this area to woodland block size classes, both in terms of hectares and numbers of individual blocks, was therefore established. It is true that these areas and numbers of blocks still had to be adjusted to cater for any inaccurracies in representation on the maps but these could be dealt with at the later sampling stage.

Work was therefore put in hand to digitise the 190 sheets which cover mainland Britain and some of the islands. The Orkney, Shetland and Western Isles in Scotland, the Isle of Man and the Isles of Scilly were totally excluded from the digitising process as were some small islands off the west coast of Scotland.

Consideration was also given at this stage to the extent to which aerial photography could be used, as it was appreciated that there were a number of functions that air photographs could fulfil. The first was that if the cover was reasonably up-to-date it could assist greatly in determining boundaries, not only of the perimeter of the woodland blocks, thereby ensuring a check on the boundary depicted on the 1 : 50 000 maps, but also internal crop boundaries indicating changes in forest type, species, age, growth rate, density of stocking and so on. The determination of internal crop boundaries without the aid of such photography can often be very time consuming. The second was that photography can also be used to determine the location and numbers of individual trees in hedgerows, in fields, as well as their occurrence in clumps or lines. Study of the photograph of an area can therefore reveal to the surveyor a great deal of the type and nature of the crops and the complexity of assessment which is going to be encountered even before entering the area; it also helps during the survey and ensures that no woodland blocks or trees are inadvertently missed.

Obviously if the photography is to yield its full potential it should be as up-to-date as possible but of course the practicalities of the situation are that photography which is less than two years old will only be available for a relatively small part of the country. In this sort of survey almost any aerial photography cover is better than none and where change is known or likely to be minimal the photography can be useful up to and beyond 10 years of age but in areas of fairly rapid change it becomes of progressively less value once it is older than five years.

Information about aerial photographic cover was then held by three major Government libraries but with the limitations that had been placed on its selection in terms of the age of the cover, and the fact that the Census had to deal with every county in England and Wales and every Region in Scotland, often with choice of differing scales and seasons, it was soon apparent that it would be easier to set up a separate register of photography. This would only include those flights that were within the selected timescale, which included a sufficiently large area of land, was of an appropriate scale, and flown at a suitable time of year. Although primarily designed to meet the needs of Census this register had other considerable advantages for the Forestry Commission's own continuing survey programme and once established could be maintained and up-dated as a record of available cover.

All the major air survey organisations were therefore contacted to supply the required flight information. This resulted in an air survey register which generally related to cover taken from 1977 onwards but in some cases information was also available for earlier periods simply by virtue of the way the records are kept. This data collection was again a substantial workload but the information was built up during the period 1978 and 1979 with arrangements being made with the major air survey companies for the records to be updated annually with those flights that the companies knew would be of interest to us. The register therefore showed at a glance whether a particular part of the country had recent cover of a suitable scale. If not, and a further search through conventional sources for older photography was unsuccessful, arrangements were made where practicable for these areas to be flown specially using the Forestry Commission's own plane or for the area to be linked in with flights of adjoining woodlands being undertaken for the Forestry Commission by the air survey companies.

The third general aspect to which thought had to be given was how the data would be collected and processed. This is an activity which is all too often neglected but when the amount of detail collected in a national survey is considered it will be evident that the collection, transmission, validation and processing of the information is a major part of the operation. The various aspects of this work are dealt with more fully in a subsequent chapter, and the statistical basis is set out in a separate publication, but certain points can be made here. The first is that the procedures used in the past for this particular part of the operation are seldom relevant at a later date. Technology in this field has moved very rapidly and the experience of the past is not particularly helpful except to highlight the massive volume of data that has to be collected, a fact not always appreciated. Secondly no matter what system is selected there are bound to be errors from various causes at various stages of the operation. It is a question of keeping this error rate within acceptable limits and as low as possible, ensuring that any errors or inconsistencies are cleared as soon as possible after the event while the knowledge is still reasonably fresh in the minds of the field teams or where checking the data is still a practical solution.

In 1947 the data were collected and recorded on specially designed field forms, the details then transferred at a later stage to punch cards and the data sorted, listed and tabulated on Hollerith machines. The transfer of the information to punch cards proved to be a major bottleneck and so in 1965 efforts were directed to optical scanning using the Lector system to cut out this stage of the operation. The error rate of this system was not high but the procedures needed to trace and amend any errors proved very time consuming. In addition, difficulties were experienced in keeping the special forms dry during the outdoor operation and also achieving markings of suitable density on the forms. Attention was therefore directed in the 1980 Census towards using a portable data terminal which could be used in the field and therefore needed to be relatively lightweight and waterproof. After examination of a number of possible machines the choice fell on the MSI – a machine with a memory of 32 K bytes – which could store the field work information obtained from at least one day's work. The details could then be transferred to a printer which meant that a hard copy of the data could be retained by the surveyor. The data were then transmitted over the public telephone system using an acoustic coupler, and stored at the receiving end on a microcomputer for subsequent validation and processing. The format of the memory in the MSI was such that the data input was made as a series of lines of data and there were limitations on the number of characters which could be entered on any one line; this meant that the input of the data had to be structured fairly carefully. However, once this had been achieved and the surveyors had become familiar with the system the procedures worked well apart from occasional difficulties with data corruption during telephonic transmission.

Although some thought could be given before the survey started as to how the data were going to be processed there was very little work that could be undertaken until decisions had been made on the nature of the data to be collected and how the final estimates for each feature (area, numbers, volume, etc.) were to be presented. Experience from the past gave a fairly clear idea of what would constitute the main tabular statements. There were, however, many other tables which could be considered but most were rejected as being of local rather than national interest. This is not to say that such items of information cannot be abstracted from the database but rather, when a national survey is being designed, the tables need to be of a standard layout so that information for one part of the country can be directly added to that from another. A series of specimen table layouts had been prepared and comments received from a wide variety of organisations including County and Regional Councils, Timber Merchants' Associations, major industrial consumers, Growers' Associations and so on. However, programmers and statisticians could not settle the design of the system until final decisions had been

made on the data to be collected, their subsequent calculation, and the presentation of the estimates including such indicators as Standard Error. This meant of course that a very considerable amount of data was collected before the statistical and other procedures had been finalised and that in some cases additional sampling had to be undertaken where insufficient samples were subsequently shown to have been selected.

Recruitment of the surveyors took place in the autumn of 1978 and, whilst preference would have been given to people with a forestry background, especially those looking for a job that would provide them with some field experience, the timing of the intake, for reasons that were largely dictated by other events, was fairly late in the year. Consequently most of the students graduating from the forestry Universities and Colleges that year were already in some form of employment. Attention was therefore directed to people in allied disciplines and most of the initial recruits were geography graduates. This might appear at first sight to have been a rather curious decision but had it not been adopted it would have meant a delay to the start of the Census until at least the following June and also, the geography graduates' training in geology, soils, land use, etc., meant that they could quite quickly be taught the forestry rudiments required. Although the training period was somewhat longer than if forestry graduates had been employed their willing and conscientious approach to the new disciplines made them very competent and efficient at the task of data collection. Numbers varied but at the outset there were 13 and the winter of 1978/79 was largely given over to training the surveyors in the new procedures both for woodlands and for non-woodland trees and familiarisation with the types of crops they were likely to encounter.

During that winter the Forestry Commission was asked by the Board of Forestry, Lands and Mines of the Isle of Man to undertake a survey of the woodlands in State ownership and this survey in March 1979 proved a useful forerunner to the Census itself. The procedures used differed from those to be adopted in the Census only in relatively minor detail and the survey gave the surveyors a good opportunity to practise crop subdivision both on aerial photographs and on the ground, to describe crop characteristics and also to use the procedures for establishing and measuring temporary sample plots. Work in the Isle of Man occupied the surveyors for a fortnight with a small rearguard remaining a further week to complete the data collection.

On their return it was decided to switch all the surveyors to the pilot survey of non-woodland trees. This is described in more detail in Chapter 6 but essentially it was undertaken because further tests had indicated that the use of soil stratification was likely to reduce the variability in the non-woodland tree count and also that 'clustering', i.e. grouping the samples, would be a more cost effective system than distributing the individual samples at random. This work largely took up the time between April and August 1979.

It will be seen that a great deal of background work had to be completed, or at least had to be well underway, before field work could start and, in addition a fair measure of co-ordination of this preparatory work was also required so that efforts were concentrated on those parts of the country where work was likely to commence. Although a considerable amount of survey work had already been done in West Sussex, it was decided to start the survey in Berkshire which was:

a. close to the Research Station;

b. smaller in terms of woodland area than any other county in south-east England and so would more quickly provide a complete set of data on which the programmers and statisticians could test their methodology;

c. known to contain a wide variety of forest crops which would test the field procedures designed during the West Sussex trial survey but not as yet finalised;

d. had recent aerial photography and had already been hand digitised on a trial basis, as contractual arrangements had not yet been entered into with Laser-Scan Laboratories.

Arrangements were therefore made to lift the Dedicated and Approved woodland data for this county from files, as that data base was still incomplete, and to draw the record of Forestry Commission woodlands in the county from the Commission's own data base which was already in being.

Work started in Berkshire while the pilot survey was still being completed and teams were gradually introduced into the county as they completed their other duties. At this stage consideration was being given to the future distribution of teams and it was decided that the most effective procedure would be for teams to work from recognised centres until such time as daily travel became excessive. The team centres were generally areas where the Field Survey Branch of the Forestry Commission already had office accommodation and where facilities such as light tables, telephones, etc., were already available. The centres adopted initially were Chester (Cheshire), Thetford (Norfolk), Hawkhurst (Kent), Alice Holt (Hampshire), Dean (Gloucester), Exeter (Devon), and Sherwood (Nottingham) and the teams were outstationed from about January 1980. The original intention was that the Field Survey staff would assist the Census teams in their work not only in providing facilities but also by helping to the ex-

tent of about 20 per cent of their time in order to speed up progress. However, this latter arrangement proved somewhat impractical and the major part of the work from these and subsequent centres was done by the Census staff alone. The only exception to this was in Wales, North, East and parts of West Scotland where because of the scattered distribution of the samples it was considered easier to involve resident Field Survey staff and to train them specifically in Census techniques.

One aspect which needs to be mentioned at this stage is the setting up by the Government of a series of studies of which one, that on the Review of Statistical Services undertaken by Lord (then Sir Derek) Rayner had a considerable impact. Census surveys are, of course, the major source of statistical information on woodlands and trees in Great Britain and consequently came under scrutiny. Alternative methods of data collection and differing intensities of sampling were both investigated but as the Census was already underway in England by this time there was little that could be done to alter the immediate arrangements. Consideration was, however, given to whether the intensity of sampling in Wales and Scotland should be reviewed in the light of the recommendations. With the arrangements already made for digitising the maps, purchasing photography, etc., a substantial amount of money had already been spent and after considerable discussion it was decided that the woodland survey should proceed in England on a county basis as planned with the exception that Durham, Tyne and Wear and Cleveland would be assessed as a County group rather than as individual counties.

In Wales the sampling frame had initially been designed to allow some of the larger Districts to be treated as units in their own right. The reasons for so doing were that the boundaries of these Districts were often similar to those of counties which existed prior to local government reorganisation, and so would have simplified the comparison of present woodland results with those of previous surveys, and also because of the relatively large size of some of the new counties. This more detailed arrangement had now to be somewhat altered in favour of surveying whole counties with the exception of West, Mid and South Glamorgan, which because of their relatively small land areas, were to be combined and sampled as one.

In Scotland the sampling frame would no longer be the District or groups of Districts but the Forestry Commission Conservancy. The lowering of the sampling intensity in Scotland was perhaps not as serious as it might at first appear as a substantial part of the woodland resource is made up of Forestry Commission and Dedicated and Approved woodland so that the area which remained to be sampled was relatively small compared with that of England or Wales.

Similar reductions were made in the levels of sampling for the non-woodland tree populations in all three countries, from those levels originally designed.

CHAPTER 5

Sample selection – woodlands

It is convenient at this stage to look at the procedures which were adopted for the selection of woodland samples for a county in England and Wales, or a Conservancy in Scotland. The discussion here is confined to 'Other' private woodlands as the methods used to obtain the requisite information for Forestry Commission and Dedicated and Approved woodlands are dealt with separately in Chapter 9.

The digitising of the woodland blocks on the 1: 50 000 sheets of each county provided for each individual woodland block a serial number, a centroid grid reference, and its area in hectares. In England these areas were recorded correct to two decimal places but it was realised that this was unnecessarily precise and for Scotland and Wales the area was recorded correct to only one place of decimals. The data were then transferred on to the Research Division's computer and sorted to provide a listing of all the woodland blocks in the county in terms of National Grid references of ascending northings within eastings. The objective was to provide a checklist of all the blocks within the county in a form whereby the location of each block and its area could be quickly and easily identified. The individual woodland areas were also allocated to classes based on block size, the selected ranges being 0.25–1.99 hectares, 2.0–9.99, 10.0–19.99, 20.0–49.99, 50.0–99.99, and 100.0 hectares and over. Any digitised blocks that were under 0.25 of a hectare in extent on the 1: 50 000 map were not further considered. If, however, any of these woodlands were subsequently found to have an actual area of 0.25 of a hectare or more they were assessed as 'Extra' woodland (see page 27). The computer was then programmed to accumulate the number of woods in each size class, to calculate their mean area and variance, and the mean size of all the woods present in the county and its variance.

The limits of the class intervals may appear somewhat arbitrary but the choices were made with three major considerations in mind. First, a separation of the area of blocks at the 2 hectares class boundary would permit direct comparison between the results of the 1980 and the 1947–49 Censuses. Secondly, it was known that the average size of woodland block

in 'Other' private ownership tended to be much smaller in area than that owned by the Forestry Commission or worked under a Dedication or Approved Woodland Scheme. Thirdly, by keeping the class limits fairly narrow at the lower end of the scale it had the effect of reducing both the number of woods and the variance in the smaller classes. The capability still exists of re-sorting the data using other class intervals should this be desired.

It needs to be appreciated, however, that all the areas derived up to this stage had been obtained from measurement of the woodland blocks as represented on the 1: 50 000 scale map and this information was subject to certain limitations:

a. maps were often out-of-date so that woods could sometimes have been established, or have been removed since the last map revision; the shape of the wood, and hence its area, could also have changed as a result of the addition of new areas of planting or colonisation, or by the clearance of trees;

b. representation errors occur on the 1: 50 000 maps whereby woods are sometimes shown slightly oversize to ensure their inclusion or slightly undersize to permit adjoining detail to be shown;

c. some areas depicted by the Ordnance Survey as woodland did not meet the basic woodland definition, or the minimum area and width constraints, adopted by the Forestry Commission for the Census, e.g. on large scale maps four densities of trees used to be shown – scattered, open, medium and close; the first two categories are not depicted at all at 1: 50 000; the others are usually shown but there can be exceptions.

Therefore it is best at this stage to consider the total woodland area derived from the 1: 50 000 map as an initial estimate which, although reasonably accurate, required a measure of updating and amendment before it could be finally adopted. The method by which these adjustments were made within an individual county was to establish the relationship between the areas of individual woodland blocks as depicted on the 1: 50 000 map and the areas of the same blocks determined from ground assessment or from recent aerial photography. However, this could only be done on a relatively small number of woods and required a decision on how best to draw an unbiased sample of such woods from within each of the size classes which would be large enough to ensure that the calculated total area for the county would be within the required precision. This level had originally been set at ±10% for 'Other' private woodland in a county but it was found that where 'Other' was the predominant ownership the subsequent addition of the unsampled areas of Forestry Commission and Dedicated and Approved woodlands to this total often did little to improve the overall precision of the estimate. A Standard Error of nearly ±10% on total area would have resulted in the associated forest type, species and age class estimates being more imprecise and so of limited value. Consequently it was decided that the level of precision should be raised.

As the number of woods, their total area, mean and variance had already been established for each size class, samples were drawn at random from within each of these classes using a constrained optimal allocation technique. This means that while the number of samples drawn from each class may be subject to restriction or weighting the process of sample selection continues until such time as the prescribed precision of the mean of all the samples combined has been attained. The major objectives involved were as follows:

    a. the precision for the estimate of total area should lie within ±5%;

    b. the number of samples to be drawn from any one class should relate to the variance of that class, i.e. those classes with high variances would require more samples than those with low ones;

    c. there should be a good representation of small blocks in the sample. Because the smaller size classes have a limited size range, they are likely to have a relatively low variance and so would normally require relatively few samples. The reason for increasing their representation was twofold; first, if there is a difference between the digitised and true areas of a block this has a much greater proportional effect on a small block than a large one and second, any regression analysis is usually assisted by having a substantial number of samples located round the overall mean of all the values. As this mean usually fell in the 2.0–9.99 hectares range it was a valid reason for ensuring that this category was well represented;

    d. there must be two or more samples to represent a population for the calculation of estimates and, where possible, a minimum of 10 was used in size classes with large populations (the smaller size classes) and a minimum of three in the classes with small populations (the larger size classes).

The procedure involved the mathematical determination of the sample sizes required paying due regard to the criteria listed above so that the resulting mean woodland block area estimated using a regression relationship would satisfy the required precision. This was done in such a way that the sample chose satisfied the conditions at minim..m cost expressed in terms of the number of blocks in each of the size classes and the likely cost of their assessment. In most counties the process resulted in about 200 woods being selected for air sampling.

Each sample wood was then located on the appropriate 1:50 000 sheet and the demarcated boundary checked with that shown on the 1:10 000 map and on the latest available aerial photograph. Any extension or reduction of the woodland block which had taken place was marked on the 1:10 000 map, the current area of the wood measured using a grid or planimeter, and the calculated area recorded together with locational details such as county, soil stratum and block number. This technique takes account of differences arising from both representational error on the 1:50 000 map and physical change in area. If, however, the sample wood was one that had also been selected for subsequent ground sampling the area calculation was deferred until the ground inspection had taken place.

The next task was to select a sub-sample of woodland blocks from those already chosen to establish total woodland area so that information could be collected on aspects such as forest type, proportions of species, age, diameter, volume, etc., which can only be collected accurately by assessment and measurement on the ground. For this sub-sample the accuracy that was sought was that the area of the predominant forest type should be correct to within ±15%.

There was already a certain amount of information available on what the predominant forest type was likely to be in each county. First, there were the results of the 1965 Census. Unfortunately these data related to all private woodlands and there was no way in which the information for 'Other' private woodlands could be separated from that for Dedicated and Approved woodlands. The overall results usually gave a fair indication of the likely dominant forest type but allowance could of course be made where it was known that the bulk of a particular forest type was in a particular ownership category. Secondly, there was information from the pilot surveys carried out in 1979 and 1981 (see page 21). With this background knowledge the means and their variance could be calculated for each size class within a county using two separate mathematical models:

a. a 'proportional model' which assumes that if a given proportion p of total woodland is assumed to be of the forest type concerned then each individual woodland block will likewise hold the same proportion p;

b. the 'Binomial model' in which each block has a probability p of being 100 per cent of that type.

The proportional model assumes that if, for example, Broadleaved High Forest was estimated to account for 70 per cent of the total type distribution in a county then it would be assumed that approximately 70 per cent of the area of each wood selected in the sample would be Broadleaved High Forest. With the Binomial model the assumption would be that seven woods out of every ten would be wholly composed of Broadleaved High Forest. Neither approach can ever reflect the real position but the Binomial model is likely to be the more realistic of the two, particularly for woods in the smaller size classes, and since it also tended to indicated a slightly higher number of samples than the proportional model it was generally accepted as the level of sampling to be adopted.

To ensure that each size class would have a minimum number of ground samples it was again decided that in any county or, in Scotland, Conservancy no class should have fewer than three samples and that the 0.25–1.99 hectare class, because of the high number of blocks involved, should never have fewer than 10. In those cases where one size class contained fewer than three woods it was combined with those in the adjoining class to produce an enlarged sampling frame. The distribution of ground samples among the size classes was then determined using a constrained optimal allocation technique similar to that used for establishing total area. Any counties which had above average levels of sampling were examined carefully to see if the reason could be established and what sort of sampling levels could be justified. Such cases were however uncommon and in general the number of samples selected for ground assessment in each county ranged between 60 and 80.

CHAPTER 6

Sample selection – non-woodland trees

The non-woodland tree survey was designed primarily to collect information about isolated trees and trees in woods under 0.25 of a hectare but its scope was extended to include other aspects some of which related to woodland of 0.25 of a hectare and over. The main features assessed in this part of the survey can be summarised as follows:

    a. the numbers, species, volume, health and location of isolated trees;

    b. the area of clumps and linear features, and also the species, volume and health of trees within them;

    c. the type and location of boundaries containing trees;

    d. any 'Extra' woodland not included in the survey of woods of 0.25 of a hectare and over.

The measurement of a resource such as isolated trees and trees in avenues or small copses has to be dealt with in a rather different fashion from that of woodland. Woodland information, as has already been discussed, was available in the form of map detail which provided a preliminary estimate of area which could then be refined to produce an updated figure of acceptable precision. In the case of non-woodland tree populations their scattered nature results in their omission from all but the largest scale maps and any sampling method has to use units of land area within which numbers of trees and their characteristics can be measured. Because it is virtually impossible to separate land area from water area in these samples, especially when the latter is in the form of small streams and ditches, the samples have to be considered as constituting surface area and subsequent calculations for the non-woodland tree assessment use land plus inland water rather than land area alone.

When field trials were first undertaken in 1976–77 it was decided that the sample unit should be based on the National Grid and its size and shape should be one kilometre square. It will be appreciated, however, that where non-woodland tree densities are high a unit of this size can contain literally thousands of trees and consequently could be extremely time-consuming to assess. The size of the unit was therefore reduced before the start of the main Census to a square 500 m × 500 m or 25 hectares. Even with this size of unit the costs of data collection were high and it was decided that once the survey was underway an assessment would be made of the time and costs involved in measuring these units and the situation reviewed. However, the problem was not solely one of the total number of trees to be counted and assessed in each sample unit. The distribution of trees in the landscape is such that the numbers found in the units could range from nil to thousands and the sampling system could undoubtedly be made much more efficient if this variability could be reduced.

One way of reducing variability is to adopt some form of stratification, i.e. subdividing the population into strata or groups whereby like values are combined in such a way that the variability within each of these strata is less than in the population as a whole. By reducing the variability in this fashion fewer samples are required to produce an overall estimate of given precision. In the case of non-woodland trees there is obviously some correlation between the numbers of trees to be found in a region and the nature of the land surface upon which they are situated, e.g. areas of high agricultural value tend to have fewer trees than those of lower value, rural areas to have fewer trees than urban areas, arable farming areas fewer than those in dairying areas, high ground fewer than low ground and so on. Some correlation therefore obviously exists between tree numbers and soil and land use over a substantial part of the country and whilst this correlation has perhaps become a little blurred of recent years owing to fairly rapid changes in agricultural practices it was clearly an obvious line to follow in the search for a way to reduce sampling intensity. Advice was therefore sought from the Forestry Commission's own Site Studies Branch and from the Soil Survey of England and Wales on the possibility of using broad site type as a basis for stratifying tree density.

The Soil Survey of England and Wales recognises 71 major soil types and obviously if all of these units were to be used many of the benefits of stratification would be lost. After discussion it was agreed that the 71 soil types could be amalgamated into soil groups but that within some of these groups consideration would need to be given to Potential Soil Moisture Deficit (PSMD). For example, a soil type in a dry area would be likely to have rather different characteristics from the same soil type in a relatively wetter area so that the land use, and consequently the tree density, would also be likely to differ. In effect three levels of PSMD were recognised – over 150 mm which was considered 'dry' and comprised those counties in England south and east of a line drawn down through the counties of Nottingham, Leicester, Northampton, Oxford, Berkshire and Hampshire; 100–150 mm PSMD which was considered 'intermediate', and less than 100 mm PSMD which included the wet uplands. In all 16 soil groups were finally selected to cover England and Wales. Urban areas have not been allocated soil types by the Soil Survey and so it was necessary to introduce a seventeenth group to cater for areas for which no published soil information was available.

For Scotland advice was sought from the Macaulay Institute for Soil Research and after discussion with them it was felt that while some of the soil groups adopted in England and Wales would be equally relevant to the Scottish situation it would be necessary in some instances to have categories that were specific to Scotland. Consequently seven additional groups were recognised resulting in a total of 24 soil groups for Great Britain.

The combination of soil types into soil groups obviously provides a framework within which the variability of the tree population could be reduced but before one can effectively draw a sample of areas from these soil groups and be sure that their numbers will be adequate to provide estimates of the required precision, it is essential to have some idea of the variability that is going to be encountered in the tree population in each of these soil groups. Theoretically it would have been possible within each county to have sampled each soil group at an assumed minimum level and then tested the precision of the answers obtained. If, as was likely, the desired precision had not been attained further sampling would have to be undertaken until the required level was reached. In practice, however, this approach is not only inefficient, in that if the sampling is done in a random and unbiased fashion it is likely to involve re-visiting the same general locations, it would also involve substantial delays arising from the need to acquire the necessary maps and aerial photographs covering the additional sites.

It was therefore decided to carry out a pilot survey to pre-determine the likely variability of the tree population in the various soil groups throughout the country; this work was carried out in England and Wales in the summer of 1979 and in Scotland just prior to the start of the main Census in 1981. In addition to providing information on the variability of the non-woodland tree population it was seen to have other potential benefits which included collecting information about the relative importance of tree species on individual soil groups, and also, as has already been discussed, obtaining details of the predominant forest type in woodland of 0.25 of a hectare and over to assist in establishing sampling levels in that category. The data were also analysed to see if it was practicable to estimate volumes of isolated trees, clumps and linear features from aerial photographs.

As the purpose of the pilot survey was to obtain generalised rather than detailed information, and because soil groups are normally independent of county boundaries, it was decided to obtain information in England and Wales on the basis of groups of counties. These county groups numbered 20 for England and Wales and reflected, where practicable, similarity of land use and of tree growth; in Scotland the Conservancy was considered to be the equivalent of the county group.

Within each county group three clusters each of six samples were sited at random in each soil group with the sample unit being 500 m × 500 m in size and thus a quarter of a square kilometre in area. Each cluster was normally 500 m wide by 3 km long. Clustering had the effect of concentrating the samples so that not only was travel and assessment time reduced but also the amount of aerial cover to be purchased or flown. On one sample unit in each cluster the surveyor assessed and recorded information for all isolated trees on species, diameter at breast height, total height, timber height in the case of broadleaves, crown diameter, health and location. Similar data were recorded for trees in clumps and linear features with the exception of crown diameter, health and location but additionally information was collected on the dimensions and area of the features. Information was also recorded on the forest type of any woodlands of 0.25 of a hectare or over which occurred in this square. On the other five sample units in the cluster only the number of isolated trees in each square was counted and no other details were taken.

The pilot survey data were analysed to obtain estimates of the relationships of crown diameter/diameter at breast height and crown diameter/volume. A satisfactory precision was obtained for isolated trees within county groups in both cases. The analyses also showed that some amalgamation of county groups could be made but it was decided not to simplify the procedure further. The range of percentage variations accounted for by crown diameter/volume relationship for conifers was between 39 and 87 per cent and for broadleaved species between 34 and 84 per cent.

For clumps and linear features similar relationships between area and volume were calculated but the estimates were not sufficiently precise unless an additional factor of height was built into the equations. As measurement of tree height from aerial photographs under field conditions is slow, and in the case of small scale photography often inaccurate, it was decided to base the volume measurements of both clumps and linear features solely on ground data.

The detail obtained from the pilot survey provided information which not only gave a measure of the variability in each of the soil groups but also indicated that in some county groups several of the soil groups might themselves be combined by virtue of the fact that their mean number of trees and the variances were similar. Consequently the large number of soil units originally recognised by the Soil Survey authorities were combined into 24 soil groups for Great Britain and these groups were in turn further combined, where appropriate, into sampling strata.

This approach obviously reduced the levels of sampling needed in individual counties but before the subject of sample numbers could be considered it was essential to set limits on the precision of the estimate that was being sought. The non-woodland survey was based on a double sampling scheme, i.e. the first stage involving the assessment of all the sampling units using aerial photographs and then the second stage assessment of checking a proportion of these on the ground to establish the 'correction factor' to be applied to the aerial photography interpretation results.

It was therefore essential that the precision of the estimate be attached to a variable which could be identified easily both from the air and on the ground. Of the three categories of isolated trees, clumps and linear features only isolated trees can be identified in terms of tree numbers with any degree of certainty. The reason is that with both clumps and linear features the tree crowns often merge on the photograph and the identification and counting of individuals can be extremely difficult. Accordingly it was decided that the standard error on the number of measurable isolated trees in a county in England and Wales, or a Conservancy in Scotland, should not exceed ±25% and that a standard error limit of ±30% should be set for the number of trees of the most widely represented species of isolated tree in the county or Conservancy. These percentages may appear unduly large but the very variable nature of the resource is such that even with soil stratification the levels of sampling required to attain these precisions during the main survey were unacceptably high in a few cases.

The results of the pilot survey were analysed to find within each county group the mean number of isolated trees per unit area on each soil group together with its associated variance and also to identify the groups which did not differ significantly in their means or variances and so could be further combined. The number of samples then required to obtain results at the required precision for individual counties within a county group was calculated using an optimal allocation technique similar to that used in the woodland assessment. The same procedure was simultaneously applied to the three most important species of isolated tree. In practice, to ensure that there would be an adequate spread of samples throughout a county, it was laid down that there would never be less than four sample clusters in any one soil stratum and never fewer than 20 samples in all. Clustering of the samples in a manner similar to the system used on the pilot survey was also adopted for the main survey and samples were again combined into groups of six forming continuous strips which were 0.5 km wide and 3.0 km long. All six samples were assessed using aerial photographs and two of the six, selected at random, were visited and assessed on the ground. The clustering of the samples in this way reduced considerably both the costs of photography and of flying as not only was the number of sites to be flown reduced, but also each strip of six samples could be flown in one run thus reducing the number of passes required. The compact nature of the sample clusters also resulted in ground survey costs being significantly reduced.

A further review of costs and progress was made following the completion of field work in the first six counties (Berkshire, Devon, Humberside, Kent, Norfolk and Merseyside). It was evident from the relatively slow progress of this aspect of the survey, and the requirements of the Rayner Committee, that further reductions in the workload would need to be made and this was effected by reducing the dimensions of each sampling unit to measure 250 m × 250 m. These smaller units of 6.25 hectares each were amalgamated into strips or clusters which still retained the original length of 3 km. Although the number of individual units in a cluster had been raised from six to twelve it was still considered statistically acceptable to ground assess only two of them. Consequently not only was the sample area to be assessed reduced to half its previous size but the ground survey area was reduced to a quarter.

The smaller size and clustering of sample units was also subsequently adopted in Scotland. However, a shortage of readily available aerial photographs of suitable age and scale in some areas and on occasion a virtual complete absence of photography would have involved a large flying programme. This had to be ruled out on the grounds of both time and cost and the following procedures were adopted instead:

a. where no aerial cover existed, was too old to be of practical use, or was small scale photography (usually smaller than 1:25 000) each cluster was visited and assessed fully for isolated tree numbers and areas of clumps and linear features, i.e. all samples in the cluster were ground visited and assessed for features that would normally have been derived from aerial photographs. Full measurements were taken in the two nominated ground samples;

b. for the remaining samples in Scotland the two stage sample based on aerial photographic interpretation backed by ground assessment on two of them was followed.

Very shortly after the start of field work in Scotland it was found that some modification of the sample size and shape used hitherto was desirable because the rapid changes of terrain, and consequently soil type, resulted in clusters regularly crossing sample strata boundaries. This occurred even though many of the additional soil types in Scotland had been classed as complexes in order to enlarge the areas of the soil groups. It was therefore decided to make the clusters more compact and although the size of the sample unit remained unchanged at 250 m × 250 m they were arranged in two rows of four samples, i.e. a block 1 km long by 500 m wide. This re-arrangement of the cluster was a considerable help to the surveyors when they were doing the work on the ground as the more compact shape reduced the amount of walking and climbing and, in those cases where only two of the eight samples had to be assessed, it limited the distance to be travelled between them. Because the total number of sample units in any one cluster had been reduced from twelve to eight, consequently reducing the area of each sample cluster from 75 to 50 hectares, the total number of aerial photographic interpretation clusters selected within the area to be sampled, in this case a Conservancy, was increased and was never fewer than 90; as was the case in England and Wales there were also never fewer than four samples per soil stratum. The choice of 90 was fairly arbitrary, being the number appropriate to about three English counties, but it needs to be remembered that non-woodland trees in Scotland are much less significant in terms of numbers than elsewhere in Great Britain and consequently a lower sampling intensity was acceptable.

Having decided upon the number of clusters required in each county by applying the optimal allocation technique, the next process was to locate the clusters at random within the various soil strata present. Because the clusters were based on the National Grid, each sampling frame, whether it was a county in England and Wales or a Conservancy in Scotland, had first to be presented in National Grid format. This was effected first by 'regularising' the county boundary on the 1:50 000 scale map, i.e. defining and marking the boundary as a series of straight lines following those kilometre grid lines which most closely approximated to the actual boundary marked on the map. Within this framework the boundaries of the soil strata were similarly plotted to divide the county into a series of blocks whose area was a whole number of square kilometres. Obviously a certain amount of rationalisation of the soil groups ensued where very small units occurred but the boundaries of the soil groups conformed as closely as possible with those agreed by the Soil Survey organisations. Using the regularised county and soil group boundaries a computer map was drawn up comprising a grid of squares, each one equivalent to a quarter of a kilometre at the scale in question. The boundaries of the sampling strata were then drawn on the computer map combining soil groups as indicated by the results of the pilot survey. The number of squares within each sample stratum was then counted and divided by the number of sampling units in the cluster to give the total number of potential clusters. The required numbers of samples were then chosen at random and located within each stratum.

Four maps were then marked up:

a. a map at a scale of 1:100 000 to show the location of all hedgerow clusters for the county or Conservancy. This was forwarded to the Forestry Commission's Air Survey Section to order any commercial photography of suitable date and scale;

b. a map at 1:50 000 scale to show targets to be flown by the Forestry Commission in those areas where no suitable commercial photography was found to exist;

c. a map at a scale of 1:50 000 to show the location of all the non-woodland targets in each county or Conservancy. This map was produced for the use of the surveyors and usually also showed the regularised county and soil strata boundaries;

d. each target was marked on to a 1:10 000 scale map. All blocks of woodland shown green on the 1:50 000 map and falling within the cluster boundaries were marked with a green verge on the 1:10 000 sheet to indicate that they had already been considered and included in the woodland survey and consequently had to be disregarded in the non-woodland assessment or in any survey of 'Extra' woodland.

CHAPTER 7

Air and ground assessment

Each field team was given as much background information as possible about the counties they were to survey and the surveyors themselves had the task of transferring information about the samples from the computer outputs to the field maps. This provided some work during periods of inclement weather and also made field staff familiar with the location of both the woodland and non-woodland samples so that they could give some thought to the planning of the work.

At the start of a county a field team would have in its possession the following maps and information:

a. a complete list of all the digitised blocks in 'Other' private ownership;

b. a summary of the above list showing the number and area of woods in each of the six woodland block sizes, their total, and their means and variances;

c. a computer produced map at 1:50 000 scale showing the boundaries and the locations of all the digitised woodland blocks (the Check Plate);

d. a list of the sample woods to be assessed in each size class and the nature of the survey, i.e. whether they were to be surveyed using aerial photographic interpretation alone or a combination of air and ground assessment. In order to allow for the possibility that some samples might need to be replaced the list contained rather more samples than were actually needed;

e. a map at 1:100 000 scale showing regularised county and soil group boundaries;

f. a map, also at 1:100 000, showing regularised strata boundaries and the locations of the non-woodland clusters;

g. a map, or series of maps, showing the flight lines of the aerial sorties involved and the air photo print numbers supplied;

h. a full set of 1:50 000 maps for the county and 1:10 000 or 1:10 560 sheets covering all the sample areas, both woodland and non-woodland. The regularised boundaries of any special land categories such as National Parks or Areas of Outstanding Natural Beauty were also marked on to the 1:50 000 maps.

Most of the above information had been collected at the Research Station at Alice Holt Lodge and the first task of the surveyors was to mark up all the information on to the maps in readiness for field work.

The boundaries of all woods selected for sampling were marked on both the 1:50 000 and 1:10 000 sheets either with a yellow verge, indicating that the wood had been selected for both aerial and ground assessment, or with an orange verge indicating that it had only been selected for aerial assessment. To ensure that all the selected woods actually did belong in 'Other' private ownership, and that no errors in recording ownership had been made at the digitising stage, the check plates were compared with maps at 1:50 000 on which all woodland blocks owned by the Forestry Commission were outlined in red and all woodlands working to a Dedicated and Approved Woodland Plan were outlined in blue. This not only located cases where there had been misidentification of ownership, or where blocks had been missed, but also indicated, in the case of the non-woodland samples, that ownership of the land could be fairly easily established.

A similar colour coding of yellow and orange to distinguish the air and ground non-woodland tree squares was also adopted with the actual outline of the strips having first been made in black. Any blocks of woodland on the 1:10 000 map which fell within the sample strips, and also appeared on the 1:50 000 map were marked with a green internal verge to indicate that they had already been included in the woodland survey and so did not need further consideration.

These copies of the 1:50 000 maps, which by now had virtually all the relevant information plotted on them, became the basis for discussion with outside organisations. The first and most obvious task was to obtain as much information as possible about the ownership of the sample woodlands and also about any recent changes that had taken place in the county in respect of both woodland area and numbers of non-woodland trees. Two sources produced the bulk of the information on these points. The first was the local Forestry Commission staff who often held information about ownership either from local knowledge or through advisory visits to estate owners and farmers. The other useful source was the local authority. In some cases there was a Forestry Officer who was able not only to provide useful details on names and addresses but also to pinpoint areas where recent tree planting had been undertaken on motorways or main trunk roads, in public parks, and so on. Where there was no Forestry Officer one of the Planning Officers could often provide the appropriate information. Even if they were not always able to confirm ownership of a property they were usually able to at least suggest an initial contact.

Other organisations such as officials of the Ministry of Agriculture, Fisheries and Food, Department of Agriculture and Fisheries for Scotland, Welsh Office Agriculture Department, National Farmers' Union, Landowners' Associations and Federations, etc., were also contacted not only as part of the general search for information about ownership but also to acquaint them of the fact that fieldwork on the Census was imminent. This personal approach provided them with an explanation of the aims and methods of the survey and gave an opportunity for the surveyor to discuss some of the problems with them. The fact that this prior approach had been made often helped to allay the concern of landowners and farmers about the survey. Information about the Census was also publicised, particularly at the outset of the survey, by radio and television. The national and local press were also informed to ensure wide publicity, and a number of organisations also mentioned the survey in newsletters to their members.

Nothing is more likely to incur natural indignation in a landowner than unauthorised entry and surveyors were expressly instructed to seek permission at all times if land had to be crossed. The identification of land and woodland owners is often quite difficult, and consequently can be very time consuming, but once the surveyor had established the names and addresses of as many as possible of the owners of the woodland blocks they sent letters giving the reason for the contact and requesting permission for entry. Because of the very scattered nature of the non-woodland tree resource, and consequently the large numbers of owners involved, contacts were not generally made on this aspect of the survey until the day of assessment when owners were asked directly for permission.

While the surveyors awaited replies to the first of the letters they continued with the non-woodland tree assessments, or with aerial photograph interpretation. The timing of this latter aspect was not critical and could be undertaken during spells of bad weather, as a break from fieldwork, or when one member of the team was absent. All surveyors were

trained in aerial photograph interpretation to the required standard at the start of the survey and each team had its own binocular mirror stereoscope to facilitate the work.

Aerial photographs formed the first stage of the sampling system in the case of both the woodland and the non-woodland surveys and the procedures adopted are discussed in the following sections.

Woodland

The woodland samples had already been marked on the appropriate 1 : 10 000 map and the boundaries of these blocks were then compared with those on the latest available aerial photographs. The photographs were used therefore to confirm that there had been no change, or alternatively to amend the boundaries shown on the map. Any extensions of the woodland block area shown by the photographs to have taken place were marked on the 1 : 10 000 map and given a green hatching and any reductions were denoted by black hatching. If lack of map detail made it difficult to define the limits of the extension or reduction in area then proportional dividers, adjusted to the scales of the map and photograph, were used to transfer the information. Once the boundaries had been confirmed the current area of the wood was measured on the 1 : 10 000 sheet using either a planimeter or hectare grid. This information was then recorded with details of the county, soil stratum and woodland block number. If the block had also been selected as a ground sample the aerial photographic interpretation was extended to include the demarcation on the map in pencil of the boundaries of the different crop types, species, or ages insofar as it was possible to determine them. Also transferred were any roads or rides not marked on the maps but which would obviously help the surveyors to locate themselves on the ground. This preparatory work saved time at the subsequent field data collection stage but the surveyor still took the aerial photographs on the ground visit as a supplementary aid.

As soon as some replies to the letters seeking entry to woodland had been received a start was made on the ground survey. Generally the surveyors were encouraged to work systematically through the woodlands to be sampled in the county, clearing one area before proceeding to the next, but this was not always practicable due to delays in response from owners, or where owners requested that surveyors confine their visits to a particular day, week or period of the year, e.g. out of the shooting season. Owners were in general very co-operative and arrangements could usually be made for entry on a date and at a time that was mutually convenient. Cases of outright refusal to permit entry were very rare but when they occurred, or where the delay in obtaining entry was likely to seriously disrupt other commitments, the matter was referred to the Census Officer who, if he was unable to persuade the owner to change his views, arranged for an alternative sample to be selected elsewhere in the county. Fortunately there were relatively few occasions when this procedure had to be implemented because it meant that new maps and new aerial cover had to be ordered and this in itself could involve considerable delay.

Prior to the ground visit the surveyor had been working on woodland information supplied by the Ordnance Survey, i.e. woodland as marked on the 1:50 000 and 1:10 000 maps and, on occasion, it was inevitable that there would be differences of opinion between the Ordnance Survey mapping and the Forestry Commission surveyor's view as to what constituted woodland. The definition of woodland used for the Census was that an area must be at least 0.25 of a hectare in extent with a width of 20 m or more, which either was bearing trees or woody shrubs to form a crop that was at least 20 per cent stocked by area, or that had borne tree crops in the past and stumps were still in evidence. If the surveyors felt that the sample woodland block did not meet these criteria it was classed as 'Technically Disforested' and a note made to this effect so that the overall county total could be correspondingly adjusted. In other instances, of course, the surveyor would find on arrival that the wood had been cleared and the site was now a field or building; in these circumstances the area was classed as 'Disforested' and the cause recorded.

If the wood satisfied the criteria for woodland then the surveyor checked that the crop boundaries that had been transferred in pencil from the aerial photographs on to the map were still valid. The aerial photographs used could sometimes be several years old and changes could have taken place to both the external and internal boundaries during this period. A boundary check was therefore made and a note taken as to whether or not the perimeter of the wood had a physical boundary such as a fence, wall or ditch that would prevent the entry of stock. The block was then sub-divided into sub-blocks or stands, each numbered serially and normally 0.5 of a hectare or more in area, and uniform for the purposes of description as regards forest type, species, age class, height class, condition, etc. Woodlands which were less than 0.5 of a hectare in area were treated as one sub-block. Sub-blocks 0.25–0.5 of a hectare in size were sometimes used but only if the crops concerned were markedly different from those surrounding them. If a county boundary ran through a wood then it always formed one boundary of that wood. Where rides or breaks were more than 20 m wide they were mapped off as blank; if 20 m or less the ride was considered to be an integral part of the woodland area. The total area of the block was measured to the nearest 0.1 of a hectare from the 1: 10 000 sheet and the areas of the various stands in the block were then similarly calculated and summed. If there was a discrepancy the area of the largest stand was adjusted to ensure that the sum of the parts equalled the whole. The area totals were then checked by another surveyor.

Where forest operations such as planting or felling were in progress at the time of assessment the stand was classified according to its probable conditions at 31 March 1980 – the operative date of the Census; otherwise the actual state of affairs as observed at the time of the survey was recorded.

Stands were allocated to one of ten forest types as follows:

a. Coniferous High Forest
b. Broadleaved High Forest
c. Broadleaved High Forest of Coppice origin
d. Mixed High Forest
e. Mixed High Forest of Coppice origin
f. Coppice
g. Coppice with Standards
h. Scrub
i. Cleared
j. Disforested

Canopy per cent was also recorded. This was an estimate of the amount of 'sky space' occupied by canopy. Entries were made in terms of percentage values at 5 per cent intervals for all forest types except Cleared and Disforested. The canopy per cent quoted related only to the tree cover so that where shrub species also occupied the upper canopy the canopy per cent was reduced accordingly.

For High Forest crops a note was made of storey as to whether it was single, or, if two storeyed, whether upper or lower. Where no separate levels of canopy could be distinguished but more than one age class was present the crop was recorded as uneven.

The distribution of species was then recorded for all forest types except Cleared and Disforested. A maximum of five species per stand could be recognised although species proportions of less than 10 per cent were usually ignored. The woodland tree list comprised 38 named species or species groups and also five combined groups namely Other conifers, Other broadleaves, Mixed conifers, Mixed broadleaves and Ornamentals. In those cases where a shrub layer was present in a stand a note was made of the species and the per cent ground cover occupied by each. An extended list of species was used which included many that only occur as shrubs, e.g.

sallow, broom, gorse, rhododendron, etc.

Crops were allocated to one of 10 planting year classes namely P71–80, 61–70, 51–60, 41–50, 31–40, 21–30, 11–20, 01–10, 1861–1900 and Pre 1861. The age bands adopted are wider in crops planted prior to 1900 because of the difficulty of assessing age in many of these older crops. Ring counts on stumps, whorl counts, local records or local knowledge were used whenever possible to put crops into the correct P Year group whilst individual P Years were only recorded where they could be assumed to be accurate.

To provide essential data for the estimation of volume in High Forest and Standards over Coppice, measurements of top height (the average height of the 100 trees of largest diameter per hectare), the mean diameter at breast height (dbh) of these trees, and the basal area per hectare were made provided the mean diameter at breast height of these crops was 7 cm or more. Depending on the size and variability of the stand, top height plots, which varied in number from three to eight, were selected and sited at random; plot size varied from 0.01 and 0.05 of a hectare depending on the number of species present in the stand. Within each plot the height of the tree of largest girth of each of the major species was measured by hypsometer and recorded together with their appropriate diameters at breast height.

Basal area was generally measured in coniferous and mixed crops by taking a series of sweeps with a relascope at intervals throughout the stand. All broadleaved crops, and coniferous and mixed crops where ground conditions made relascope sweeps impractical, were dealt with by measuring the dbh of all trees within a sample plot of appropriate size. The reason for treating broadleaved crops differently from coniferous and mixed crops is that coniferous crops tend to be even aged and provided the species, age, basal area, and top height are known a reasonable estimate of volume and mean crop diameter can be made. Broadleaved crops however, are often variably stocked and composed of trees of more than one age, and consequently size, so that mean crop diameter and volume can seldom be deduced with accuracy from a knowledge of the species, height and basal area alone. Basal area plots measured in broadleaved crops varied in size according to stocking and the plot size selected was normally one that would include somewhere between 7 and 20 trees of each of the two major species; the number of plots selected was usually half that of the number selected for top height measurement so that in effect every second top height plot became a basal area plot. Within each plot every tree of 7 cm dbh and over was girthed using a diameter tape and its species and diameter recorded.

Non-woodland trees

The aerial photographs covering each sample cluster were identified and the boundaries of the cluster transferred from the OS 1:10 000 map on to the photographs. In some cases this could be done directly using detail common to both the print and the map but in others the scale of the photographs had first to be determined and the boundaries then transferred using proportional dividers. When this occurred the photographic scale and the proportional divider setting were marked on the reverse of the photograph. The prints were then placed under a stereoscope and the surveyor worked systematically through each individual sample unit identifying, marking on the photograph, and counting all isolated trees and measuring their crown diameters as depicted on the photograph. If there were fewer than 15 isolated trees in the square then the crown diameters of all of them were measured. If there were more than 15 trees then at least 15 of them, chosen at random, were measured for crown diameter. Similarly clumps, defined as small woods or groups of trees of less than 0.25 of a hectare in area, and linear features, which are strips of woody vegetation less than 20 m mean width, crown edge to crown edge, and more than 25 m long, were also marked on the photograph, and their width and length measured. Isolated trees were numbered in ink on the photograph with an individual point and clumps and linear features circled and annotated with a C or an L to ease identification at a later stage.

At the same time as the air photographic interpreter was recording details of isolated trees, clumps, linear features, etc., he also noted whether the trees were growing on a boundary or were open grown. In the case of clumps and linear features the presence or absence of associated boundary information in the data input indicated which category of boundary was involved, but isolated trees were specifically coded as being either on a boundary or open grown. Information on the nature of the boundary and its location were also recorded and these features are discussed in greater detail later in the Chapter.

Certain other features were also assessed at the aerial photograph interpretation stage. One of these was 'Extra' woodland, a category which has already been briefly discussed. It comprised blocks of woodland occurring in the sample cluster which were greater than 0.25 of a hectare in area but which either did not appear on the relevant 1:50 000 map or were shown as being under 0.25 of a hectare in area and consequently not included in the digitised totals. These additional areas could occur as isolated blocks, provided they were in 'Other' private ownership, or as extensions to Forestry Commission or

Dedicated or Approved woodlands provided they were not themselves in one of these ownership categories. Because these 'Extra' woodland blocks can be of variable size, and so could overlap the boundaries of the sampling unit, a convention had to be introduced whereby if this 'Extra' woodland fell partly in and partly out of a sample unit then provided the southernmost tip of the wood fell inside the square the whole wood was assessed, including that part that fell outside the sample unit. If it did not then the whole wood was disregarded. The area assessed as 'Extra' woodland in Great Britain is quite substantial – possibly as much as 80 to 90 000 hectares or over 10 per cent of the 'Other' private woodland total – and was mainly the result of woods arising, often by natural means, between the date of the last map revision and the date of the Census.

It was decided, however, not to include this woodland total in the overall estimate for three reasons. First, the somewhat arbitrary convention that had to be used to decide whether the block was to be counted or not meant that the overall estimate could be imprecise. Secondly, although 'Extra' woodland might sometimes occur in a sample square that was to be ground assessed, it more often would not, and in these latter instances the detail would need to be interpreted from aerial photographs and consequently there would be virtually no information on forest type, species, ages, etc. Thirdly, unless this information is available the 'Extra' woodland area would otherwise need to have been spread pro rata over the whole range of the 'Other' private woodland categories and it is most unlikely that this would be an appropriate way of doing it. Accordingly, it has to be recognised that there is a substantial area of woodland which is not marked on the 1: 50 000 maps, has not therefore been included in the present Census but which nevertheless is present and at least part of this area will be included in future assessments if, as seems likely, alternative methods of determining total woodland area are adopted. The virtual exclusion of this sizeable area of 'Extra' woodland from the results of the present Census is obviously unsatisfactory and ways of overcoming this problem in future are under consideration.

The next stage of the work was to visit the appropriate sample squares on the ground with the object of providing information which could then be applied as a correction factor to the results obtained by aerial photographic interpretation and secondly, to provide data on species, diameter, volume and health features which can only be obtained by ground assessment. It must be made clear at this stage that the aerial photographic and field assessments were completed and kept entirely separate and the surveyors were not permitted to amend any

of the data if they found on ground inspection that their aerial photographic interpretation had been at fault. This made it possible to aim for a standard conversion factor for each county although this factor changed as surveyors progressed through a region. However, in time, they gained from the experience of being able to compare their own air and ground results and profited from their mistakes.

Two of the squares in the cluster had been previously selected at random for ground checking and the first task was to establish whether 'Extra' woodland was present or not. If there was then its area was computed and a note made of its forest type but no other details recorded. The rest of the tree population on the two ground sample squares was then assessed systematically and classed as being of measurable size or not. For a tree to be classed as measurable it had to have a diameter at breast height of not less than 7 cm, have a persistent axis and not have been pruned in such a manner as to limit growth. Trees less than 7 cm in diameter and at least 1.5 m tall were counted but not girthed provided they had a persistent axis, were individuals rather than coppice shoots, and had obviously been selected for retention. Trees under 1.5 m in height were also counted provided they had been planted and occurred in clumps and linear features, e.g. motorway planting, shelterbelts, etc., i.e. obviously being tended and likely to grow on.

The details collected for each tree during the ground assessment were species, diameter at breast height, total height, timber height in the case of broadleaves, and health. The list of species reported upon was essentially the same as that used for woodland species. Diameter at breast height was measured by tape. Total height, provided it exceeded 5 m, was measured by hypsometer but heights lower than this could be estimated; timber height of broadleaved trees was also recorded. Broadleaved trees with a timber height of less than 2 m or any trees, whether conifer or broadleaved, suffering extensive decay or rot were considered to have no volume. There were obvious occasions when it was impossible to make direct measurements and special provision had to be made for trees in relatively inaccessible positions such as gardens, fields with growing crops, etc. The procedure was to estimate the dimensions of such trees using the measurements of nearby trees as a guide.

In general, all isolated trees had their dimensions measured or estimated but in clumps and linear features the numbers of trees involved made alternative procedures necessary. First the width and length of all clumps and linear features were measured from crown edge to crown edge on the ground so that their areas could be computed and compared with those derived from the aerial photograph and second, where these features contained 10 measurable trees or less all trees were measured for volume, but where there were more than 10 trees a minimum sample of 10 was chosen systematically from the group and the sampling faction noted. The predominant forest type of clumps and linear features was also recorded.

The location of all trees growing in the sample square was described, the first main division being as to whether a tree occurred in a rural or urban situation and secondly, whether it was open grown or on a boundary. The difficulties of dealing with the many small trees and the multiplicity of boundaries in urban situations made it necessary to consider all trees in urban areas as being open grown.

The types of boundaries containing tree growth recognised were hedges, overgrown hedges, walls, banks and other. Trees were classed as occurring on a boundary if their butts lay within 2 m of it. Eight types of locations of boundary and open grown trees were distinguished: roadside, waterside, field, garden, railside, public open space, parks and enclosed premises, the last named covering such areas as landscaped areas around factories, educational establishments, waterworks, or private parkland.

The continuing spread of Dutch elm disease, and the fact that a substantial part of the non-woodland tree resource is fairly old, highlighted the need to collect information about the general health of non-woodland trees. It will be appreciated, however, that such an assessment can only be made in general terms and, particularly in the case of broadleaved trees, the time of year at which the surveyor made his visit, could be relevant. Live trees were assessed as being in one of the three general conditions namely good health, moderate health, or poor health; dead trees were allocated to a separate category.

Symptoms of ill-health can take a variety of forms and the surveyor looked at three in particular: the state of the crown, the bole and the roots. Crown damage manifests itself usually by broken or dead branches in the upper crown, abnormally small leaves, die-back, premature discolouration or defoliation, etc. Bole damage was indicated by the bark having been removed from part of the stem, dead bark, cankers, or obvious signs of rot or decay, while root damage was shown by signs that the tree was suffering wind rock or exposure of the root plate. If none of the above symptoms was present the health of the tree was assessed as good, if one symptom was present the health was assessed as moderate, and if two or more symptoms were present the health was assessed as poor. Up to 10 per cent of dead branches in the upper crown of oak, giving rise to a stag headed appearance, was permitted before downgrading was considered but dead branches or areas of dead bark in beech automatically classified the health of the tree as poor. Elm was a rather special case because of the often rapid onset of the symptoms but the surveyor checked the state of the tree as far as possible for obvious signs of beetle emergence, dead branches, premature browning of the leaves, etc. An elm tree that was suffering severe damage and was not expected to live was placed in the dead category.

Once the general state of the trees in the county or region had been assessed it could then be compared with the life expectancy tables which give some measure of the number of years a tree of a given species and size might be expected to live. Obviously such a classification, which is based on national averages, may not apply to a particular region where the life expectancy of trees can be different. What can be done, however, is to assess the relevance of the tables to the particular region and if it is felt that the tables for example, understate life expectancy in the various classes by 10 years then the table can be adjusted accordingly. In this way it is possible to judge the number of years that trees in the local landscape can expect to survive if left undisturbed.

Monitoring results

During the course of the survey, and especially at the start, it was necessary to monitor results as closely as possible. Teams, although largely trained in one county (Berkshire) were soon working in various parts of the country and eventually as many as 10 counties were under survey at any one time. Therefore the main requirements were that the standards of classification should be consistent among all the survey parties and that, because of the relatively small sampling fraction, the samples should also be accurately assessed. The responsibility for ensuring this lay with the supervisors. It was done largely by their maintaining close contact with the surveyors, visiting as many sites as possible with them, and inspecting areas already assessed whenever practicable. Many lessons were learned, both by the surveyors and the supervisors during the first year of the survey as new crops and new combinations of forest types presented themselves. Eventually common standards were attained but it did entail going back to reassess certain crops which, in retrospect, were considered to have been wrongly classified.

Similarly, the methods of monitoring the precision of the estimates were developed over time. Initially surveyors included in their progress reports the areas of each forest type in each woodland block, and the numbers of isolated trees in each aerial or ground visit sample unit of the non-woodland survey. From these data was possible to calculate simple estimates of the mean and variance of the sample populations and by combining the results of each stratum to calculate the precision of the estimate for each county. As the survey progressed programs were written which could produce these estimates directly using the data transmitted to the computer by the surveyors.

CHAPTER 8

Data transmission and storage

All woodland Censuses carried out by the Forestry Commission have been characterised by broadly similar field methods. These can be contrasted with the complete change in methods employed to handle the data once they have been collected in the field. Advances in technology have resulted in savings in manpower at each new survey but in turn they have created difficulties and it is unfortunate in many ways that this progress has been so swift that past experience cannot always be of much help in solving present problems.

The essential requirements of the data handling systems are, however, relatively constant from one decade to the next and can be summarised as follows:

a. the system must allow easy checking of data at all stages, from the moment of field collection until it is stored at the Research Station. The quantities of data collected are so large and the structure sufficiently complex to allow large numbers of insidious errors to go undetected unless facilities and time are made available to correct them;

b. it is an advantage to be able to forward data to the Research Station as soon as possible after collection with the minimum of loss by corruption. This allows a more rapid turn-round of information and mistakes can therefore be identified more speedily;

c. all woodland Censuses have depended upon the collection of data in the field in weather conditions which are sometimes severe. Whatever technology is employed for data capture in the field, whether it be pencil and paper or microelectronics, must be capable of functioning effectively under all outdoor conditions.

d. the system, once employed, should have as few inbuilt limitations as possible. For example, it is undesirable to employ a data collection device that imposes sequential access if this means more walking in the field to collect the data in the required sequence. The requirement is, in short, a system that works for the Census surveyor rather than makes the surveyor work for it;

e. the availability of an in-house computer with

suitable capacity is an essential requirement in any Census as it is used for storing, validating and editing data, obtaining print-outs and calculating large numbers of estimates and their standard errors. An in-house system is preferable to a bureau because it allows much freer access for editing and program development.

When Census data collection was being considered portable battery operated data terminals had recently become available and it was felt that the characteristics of these machines could be advantageously employed for field data capture and transmission to the Research Station. The machine chosen was the MSI/88 with 32 K bytes of memory and a detachable acoustic coupler module. The advantages of such a machine were essentially:

a. portable terminals obviate the need for any re-entering of data thus saving both time and expense;

b. they enable data to be sent directly into a computer via an acoustic coupler, telephone and modem to a central computer. This again saves time and worry of lost data through the postal system. If desired, it could have enabled surveyors to collect data in the field and send them directly to Alice Holt Lodge via the nearest public call box;

c. the model chosen was able to store up to 32 K bytes of data – more than adequate both for a day's work and for holding data collected from one wood or one non-woodland sample square;

d. the MSI terminals were reasonably weatherproof and easier to handle than sheets of paper in wintry weather;

e. the MSI had 'programmable' transmission characteristics which meant that it could be adapted to transfer data to a wide variety of computer hardware.

The disadvantages were mainly characteristics of this particular machine:

a. only one line of 16 digits of data could be displayed on the screen at one time. This was sufficient for data entry but recording information on a piece of paper would have enabled the surveyor to review much more data at a time and verify their correctness more easily on site;

b. although data could be entered into one of several 'pages' of memory on the MSI, the method of data entry was strictly sequential. This meant that a line of omitted data could not easily be re-entered and, on occasions, data had to be entered in a way that presented difficulties to the surveyors;

c. safeguards were necessary to prevent loss or corruption of data before and during transmission. It was later found necessary to purchase a printing machine to list the data prior to transmission, so that the surveyor could retain a copy of the information, and to transmit only from an office telephone as several attempts might be necessary before the data were accurately received.

At the start of the Census a computer of suitable capacity that could be linked to a modem was not available at the Research Station. Consequently, it was decided to purchase a Transdata CX400 microcomputer with two floppy disc drives to receive the data. The machine was linked to a VDU and a Post Office telephone modem through two of its four independent communication ports.

It was found necessary to equip the CX400 with 64 K of memory to enable it to have sufficient capacity to handle its own operating system, a program to transfer data from modem to floppy disc, and up to 32 K of data from an MSI/88. The discs used were single sided 0.25 megabyte floppy discs which were sufficient to store data received in any one day, and could cater for up to a week. It was later found necessary to purchase a second CX400 of identical capacity to act as a stand-by when the first was unserviceable. This was usefully deployed elsewhere in the Research Station at other times.

As a temporary expedient a UNIVAC 1108 at the UCC bureau was employed as a means of storing data on a more permanent basis from the floppy discs. In the early part of 1980 the Research Station was equipped with a PRIME 400 mini-computer for its data handling and analysis and this was used for the Census. A CALCOMP plotter was also acquired for graphical output. The PRIME was later upgraded to a PRIME 550 and two magnetic tape drives added. A single, permanently attached 160 megabyte disc drive was made available for Census use although initially not all of this was required. However, during the later stages of the Census the large volume of data in hand meant that a considerable amount of information had to be archived on magnetic tape to relieve the demand on disc storage.

The MSI/88 is a machine designed specifically for stock control systems and is therefore suited to sequential data collection where the data structure is of a linear nature. It presented a number of prob-

lems therefore when we came to decide on how best to 'fit' a rather complex hierarchical data structure where sequential data entry is a hindrance. The format chosen was to a large extent a compromise, and hinged on the following four criteria:

a. it was necessary to ensure that each and every variable was uniquely identifiable, if necessary by using explicit recognition codes or characters;

b. it was desirable to minimise the number of data items needed to be entered;

c. it was desirable to minimise the amount of footwork the surveyors were required to do;

d. it was desirable to maximise the 'robustness' of the data, i.e. ensure that all errors or omissions were readily recognisable and that correct values were not wrongly interpreted after the occurrence of an error.

The format chosen was biased somewhat towards satisfying criteria a and b rather than c and d but fortunately it was found that short sequences of data could be fitted easily on the 16 digit line of the MSI. Thus one line of data could hold all the values recorded from one tree, or could describe one component of the canopy of a woodland stand. This had the advantage of making a short set of values visible, and since they were all on one line, the imposition of an entry sequence did not create a problem for the surveyor. An example of MSI format is given in Figure 1. Facilities had to be included in specific places to allow 'parallel' entry, e.g. in recording data for woodland it was often necessary to be able to enter data for the crop (species, age, etc.) at the same time as the volume assessment (diameters, top heights, etc.).

One disadvantage with the format adopted was that it was a little inflexible and did not allow some kinds of alteration once it was put into use. This did not turn out to be a serious problem however, and the same basic format was kept throughout the Census.

In order to achieve a smooth throughput of data from the field to the Research Station it was necessary to equip the CX400 with an automatic answering facility. This enabled surveyors to send data when they wished and to some extent relieved staff at the Research Station from having to answer frequent telephone calls.

The facility took the form of a program written in BASIC, which could answer the telephone and transfer data from the modem to the computer's memory, and then write the data into a uniquely named file on floppy disc. A certain amount of development work was necessary to ensure that the rate of transmission of data by the MSI matched the rate of reception by the program, and it was found possible to equip the 'auto-answer' program with the

DATA TRANSMISSION AND STORAGE

Surveyors' numbers and date ——————

Sample square heading ——————
  101 — County Code
  12 — Soil stratum
  45607890 — Grid reference

Trees of less than 7 cm dbh ——————
  21 — Species Code
  0002 — Number of trees
  2 — Health Code

```
=14=15=280186===1=
=14=15=280186===1=
2222============++
101=12=45607890=++
-1=01===========++
40=024=11.8=00=2++
28=016=13.9=04=1++
32=050=00.0=00=0++
33=062=20.2=06=2++
21=000=0023=00=1++
23=052=16.9=02=1++
21=000=0002=00=2++
28=017=12.6=06=1++
29=065=17.0=12=1++
-1=01===========++
35=062=22.4=05=1++
25=013=09.3=02=1++
23=052=17.6=03=1++
21=035=13.2=04=2++
26=023=11.7=03=1++
-3=01=03010=2=01++
23=024=14.2=02=1++
17=036=27.1=13=1++
-1=01=000=2=1000++
```

—————— Hedgerow ground data code

—————— Start of boundary

—————— Individual tree on boundary
  33 — Species code
  062 — Diameter at breast height
  20.2 — Tree height in metres
  06 — Timber height in metres
  2 — Health Code
—————— Boundary code repeated
—————— First tree in clump

—————— Clump details
  01 — Serial number
  03 — Width in metres
  010 — Length in metres
  2 — Forest type
  01 — Sampling fraction

End of boundary ——————
  01 — Serial number
  2 — Boundary type code
  1000 — Location code

Figure 1. Computer print-out subsequent to telephone transmission of field data collected on a portable MSI/88 terminal.

capability of sending audible tone signals down the telephone line to a surveyor transmitting data. Thus it was possible, at the start of a call, to send a signal confirming that the auto-answer was operational, at which point the surveyor would transmit. If after a suitable pause no data were received a 'weak transmission' signal was sent indicating to the surveyor to try again. One of two signals would be sent at the end of transmission, either 'data received OK' or 'error rate unacceptably high' – in which case the data had to be re-transmitted.

On occasions, interference on the telephone line could result in the transmission being terminated without the surveyor being aware that this had happened. In order to secure against data loss the surveyor could compare the number of characters held in the memory on the MSI with the number of characters received by the CX400 and re-transmit if they were not equal. This, of course, involved another telephone call to the Research Station.

In order to minimise the telephone charges transmissions were only carried out in the afternoon or evening. In the morning the CX400 was used to transfer data (still in its original MSI format) to the PRIME 550 for subsequent validation and storage.

There were two important tasks to be undertaken with the data once it had been received at the Research Station. Firstly, to identify and flag errors and secondly, to change the format and structure into that more suited as input for subsequent analysis programs. The validation program was designed to perform both of these tasks. Ideally it would have been preferable to split the two functions, since it was possible that an error in a 'recognition' field could result in valid data being incorrectly sorted. Part of the process of identifying errors, however, is in producing clear, readable print-outs and this could not be done without some sorting and reorganisation; hence, the two tasks had to be tackled together.

Essentially errors arising in data collection and transmission can be classified into four types:

a. mistakes made in taking a measurement or observation with the incorrect values registered correctly;

b. gross errors in data recording, for example registering a diameter to be 251 cm instead of 51;

c. errors in an 'identification' field. This could be, for example, a grid reference or a character identifying a variable;

d. corruption of data during transmission.

The first type of error is not normally detectable unless it is large and is always a risk no matter what method is used. Errors of the second type could be easily captured by the validation program which, after identifying the variable to which the item of data belonged, would check that it fell within a sensible predetermined range. Errors of the third type were more indirectly traceable; grid references of sampled areas were always checked later as a matter of course. Identification fields in the data format referred to a whole line of data, hence an error in this character would cause the program to misidentify a line, and since the range of values of one line type differed substantially from another an error would invariably be flagged. Errors of the fourth type were normally traceable because the corruption of data was usually quite drastic. For example, digits being altered to letters or short sequences of data being obliterated. The technique for identifying all types of error within the validation program was simply to identify the data line type and check that each data value on that line fell between an acceptable minimum and maximum. Should an error arise, or be suspected, an appropriate error code would appear adjacent to the line when the data were printed out. A few other checks were built into the program, for example comparing the relationship between one value and another.

As the data from the field were validated they were presented in the form of a readable print-out and returned to the appropriate surveyor for checking. At the same time a copy of the data was retained on the PRIME 550 in a simple sequential access file in a form that was much more readable for subsequent analysis. Each of these files was identified by a grid reference and grouped into County Directories. If any data required amendment the corrections would be marked on the print-out by the surveyor and returned to the Research Station where it would be edited into the data held on the computer. As a safeguard against loss of data during transmission the surveyor could obtain a listing of data prior to transmission using his printer which could subsequently be compared to the print-out from the Research Station. At the start of the Census an alternative method of correcting errors was attempted. This system enabled the surveyor to correct previously stored data in a subsequent transmission whereupon the validation program would implement the changes to the data file. This had the advantage of cutting out the editing of data but unfortunately involved the input of a relatively long string of digits to identify the location of the error and alter it. As a result errors in the identification sequence were frequent and therefore the error remained uncorrected, or worse, was 'corrected' incorrectly. On balance, it was found to be quicker and simpler to employ someone to edit the data at the Research Station.

CHAPTER 9

Calculation and presentation of the results – woodlands

Area estimates

The results presented in each Country, Conservancy and County Report relate to three ownership categories: Forestry Commission woodlands, Dedicated and Approved woodlands and 'Other' private woodlands. For the first two ownership classes data were already available and could be transferred, after some adjustment and amendment to ensure compatibility, from other computerised systems. The data for 'Other' private woodlands were derived from the Census. Because of the characteristics of the data and the way in which they were formatted the procedures needed to complete the final tabular statements differed for each ownership class.

The Forestry Commission sub-compartment database

This database is maintained and updated annually at Alice Holt. It lists the major crop characteristics of every individual sub-compartment under Forestry Commission management together with other information including its location in terms of local authority, county or district. The definitions of forest type, age class intervals, lists of species, used in the database were all comparable with those used in the Census and there was, in fact, much more information available in the Forestry Commission database than was actually required for Census purposes. Consequently the task was primarily one of deleting data which were not required and amalgamating the remainder to fit the rather broader classifications adopted for the Census.

There were, however, four categories where information was either deficient or missing. The first two were the Low Grade Broadleaved category where species and age had not always been recorded and Retained Scrub where species had sometimes been omitted. In these instances contact was made with the forests concerned and the supplementary data added to the database. The third category included arboreta such as Bedgebury, whose areas at that time were not in the plantation totals; in these cases the area was allocated to individual species as

far as was practicable and the balance allocated to Mixed conifers or Mixed broadleaves. Determining age was a problem in some instances but the best possible allocation was made. The fourth category was woodland which occurred on Forestry Commission land but was outside the plantations and thus outside the scope of normal forest inventory, e.g. scrub woodland on agricultural holdings. All Conservancies were contacted to see if the area concerned was likely to be significant and in general it proved not to be. In one case, however, that of the New Forest, there was a considerable area of unsurveyed woodland present on the Open Forest and this was assessed specially and subsequently added to the database totals.

Although updating of the Forestry Commission database is an annual process special efforts are made at five yearly intervals to supplement the information contained in it in readiness for the quinquennial valuations. One such year was 1980–81 and as a result the database was much more informative for that year than that for 1980, the operative year of the Census. Accordingly it was found easier to readjust the 1981 database figures back to 1980 than to add missing locational and other data to the 1980 database. The required data were therefore abstracted from the database, transferred to special files, amended, sorted and summarised as necessary so as to produce the output in the form of cell entries similar to those required for the final tables of the Census Report. The data were held in this form so that the figures could either be presented as one set of ownership totals in their own right or capable of having corresponding details added to them from the other two ownership classes.

Dedicated and Approved woodland database

If the Forestry Commission database contained rather too much information for Census needs then the PW5 database, which related to Dedicated and Approved woodlands, contained rather too little. The statement of growing stock contained in the Plan of Operations is fairly basic in its form of presentation and there was certainly no obligation on an owner to supply additional detail. For example, although it is possible to distinguish Coppice crops from Coppice with Standards on the form, and to recognise the coppice species in both cases, there is no place for the species of standard to be recorded. Also, distinction between Scrub and Felled land is not asked for and the two types were combined and presented as one total 'Other Land for Planting'; for this reason sub-division of this total for Census purposes had to be made in a quite arbitrary way by allocating entries alternately to the Scrub and Cleared categories.

Because the areas were relatively small in both these instances neither of these classification problems were particularly serious but it will also be appreciated that with the data coming from a large number of private owners the degree of detail and the accuracy of the information provided was bound to vary. For example crops with more than one species present would sometimes be sub-divided to show the constituent species and their respective areas whilst in others the crops might simply be classed as Mixed conifers or Mixed broadleaves with no indication as to predominant species. In the latter instance crops were at least capable of being allocated to the correct forest type and usually to the right age class but it does mean that it is difficult to compare change in the areas of individual species over time when the area of mixed crops is so much higher on this occasion than in previous Censuses. Requests for information about crop age also cause difficulties, but it is an aspect where there is often doubt and uncertainty especially where crops exceed 100 years of age and there are few records to help substantiate the estimate. It must be made clear, however, that the PW5 form, and the information on it, had not been designed to be used for Census purposes and although there were obvious shortcomings to some of the data the great bulk of the information was in a form in which it could be confidently accepted for use in the national inventory.

Up until 1977 the PW5 form contained in the Plan of Operations was not designed for computer input and consequently as plans came up for renewal during the period 1978–80 owners were asked to submit their data on the newly designed computer input form. Plans of Operations are normally revised on a five-year cycle and consequently owners who had submitted plans in 1976 and 1977 had not long completed supplying information on the old type forms. As the work of data abstraction had to be completed by 1980 the Commission itself undertook the task of transferring the information for these two years on to the new input documents.

The information asked for on the new form was essentially the same as that contained on the old one but owners were asked where possible to sub-divide some of the larger or more complex compartments in order to provide more detail, to give a more accurate estimate of age where they could and, where practicable to translate some of the written comments into numerical detail. This re-sorting and amplification of the data proved to be quite a time consuming operation and the efforts of private owners and agents to provide the fuller and more accurate data were very much appreciated.

The five-year revision period of the Plans did, however, create a problem in that although 'he operative date of the Census was 31 March 1980 the mean date of the Dedicated and Approved woodland data is probably nearer 1978 as a result of Plans being submitted annually over the period 1976–80. It is unlikely this will have had a major effect on the overall total of the category but it will certainly have had an effect on the P71–80 age class, especially for conifers in Scotland where the areas of new planting alone during the latter half of the decade were ranging from 5–6000 ha per annum. This particular P year class in Great Britain could therefore be deficient by some 15–18 000 hectares. However, the required information could only have come from the computerisation of the grant payment forms for the years in question; this was not a practicable solution as it would have been a major task and even if undertaken would not have supplied the necessary information on the species used.

All forms completed by owners, their agents or by Forestry Commission staff were submitted to Forestry Commission Headquarters in Edinburgh where they went through a routine checking process and were returned to the Conservancy concerned if major errors or omissions were found. In some cases a search through the files produced the required information, or local private woodland staffs were able to help from their own knowledge, but sometimes there was no option but to get the information from the estate concerned or to make an arbitrary judgement based on the best evidence. Some of the older plans, for example, had crops where the age was merely given as pre-1900. These were given a special coding and at a later stage every third block was allocated to the Pre-1861 age class and the balance to the P1861–1900 class. The areas involved were again not large but this arbitrary allocation process may possibly distort to some degree the area and species distributions in the older age classes.

The data were then processed by the Scottish Office computer and tabular statements produced showing the distribution by counties, Conservancies and countries of the various forest types and the breakdown of High Forest totals by species and age classes. These statements provided the Census Section with the basic figures against which other analyses could be run and checked. A copy of the individual sub-compartment data was also passed to the Census Section at the Research Station in the form of magnetic tape as part of the basic Census record. To safeguard against possible misuse of the data if they were ever processed out of Forestry Commission control the only ownership information on the tape is in the form of an estate code number which can only be traced by reference back to the Conservancy concerned.

This exercise also brought to light quite a number of estates in the Approved Woodland Scheme which had not renewed their plans at the last date of revision and were unlikely to be interested in participating further in the Scheme. Because these estate woodlands were still marked on the Conservancy maps as being within the Approved Woodland Scheme they were consequently denoted as such on the Census maps that were sent for digitising. However, because no PW5 was then required for such estates as they were no longer operating under the Scheme, there were consequently no area records for them in the database. Information for these estates had therefore to be obtained from the files and the data added to the final tabular statements for 'Other' private woodlands, the category to which these estates most appropriately belonged.

'Other' private woodlands

The area of 'Other' private woodland was established using regression analysis with the digitised area of each sample block being compared with its actual ground area. If the results of the regression analysis then showed, for example, that the mean of the ground surveyed blocks in the 2.0–9.99 hectares class was 0.2 of a hectare greater than the mean of the same blocks derived from digitisation then all the woods in the digitised list for that county that had areas between 1.80 and 9.79 hectares were counted and their areas totalled. These woods then formed the new 2.0–9.99 hectares size category, i.e. the class limits of the digitised data were adjusted with the extent of that adjustment being based on the regression analysis. Similar adjustments were made for each of the other five woodland block sizes.

Another change that had to be catered for was disforestation, i.e. where the sample woods no longer existed as woodland and the land had been wholly converted to other uses. The number of woods found to be disforested in any one size class were determined from the air photographs or subsequent ground sampling and expressed as a proportion of the number of woods sampled in that particular size class. A pro rata reduction was then made to the class total and the remaining number of woods in each class was then multiplied by the mean area of

the class established by the regression; the revised sum of the class totals then formed the best estimate of the total area and number of woods in 'Other' ownership in any given county.

The breakdown of the woodland totals by forest type, species, age class, etc. was also calculated separately for each woodland size class. The ground survey data from the sample woods were subdivided into their constituent sub-blocks and percentages established for each forest type and, within them, species and age class as appropriate. These percentages were then applied to the total calculated area of each of the six woodland block size classes determined above. The calculated areas of each category within each size class were then summed to give the overall pattern for the county or Conservancy. This approach meant that each woodland block size was treated as though it were a stratum and thus due weight was given to the likelihood that particular forest types or species might be differently represented in each size class. The overall results were again held as a set of standard tables so that they could be utilised on their own or added to the corresponding cell entries of the other ownership classes.

Monitoring precision targets

As mentioned in Chapter 7 methods of monitoring the precision of the estimates were developed to ensure that the predetermined degrees of accuracy for each county were attained. Initially the surveyors supplied the requisite information but later programs were written to extract the principal variables from the data and calculate the county estimates and standard errors.

Whenever possible, monitoring was continuous once 50 per cent of the samples within a county had been completed so that early notice was given of any counties where it was unlikely that the desired precision would be achieved. The lists of random samples produced at the start of each county contained more targets than the sample selection routine indicated so that if extra samples were required they could be readily provided.

It was also possible to use the monitoring results to discover any trends or bias that a particular county or survey team might have had and this helped supervisors in deciding which aspect of the survey to concentrate on and, as the results were site specific, which samples to revisit either to reassess or to use as a standard for comparison.

Precision achieved

The estimates of the total area of woodland for all counties but one lie within the desired precision of

±5%, ranging from 0.3 per cent in South Scotland Conservancy to 3.6 per cent in Surrey. In terms of 'Other' private woodland the achieved precisions lie between 1.8 per cent in West Midlands and 4.6 per cent in West Scotland Conservancy. The exception is Greater London whose precision for total woodland area is 6.8 per cent but where the costs of improving the estimate were considered to be too high in relation to the potential benefits.

The precisions of the estimate of the predominant forest type range from 0.6 per cent in South Scotland Conservancy to 13.2 per cent in Kent. Assuming that acceptable standards of measurement and assessment applied throughout, there are several reasons for both the variances and their range:

a. although the first estimate of the area of 'Other' woodland was based on complete digitisation the final estimate was derived by sampling and so has a sampling error associated with it.

b. although the most up to date versions of the woodland plates were used some of them dated back to 1963 and often many changes had taken place.

c. it was found that the rate of change in the industrial counties was rapid, even though the maps were relatively new in many instances. The major influences were housebuilding and industrial development, including spoil tips, which caused both the expected disforestation and also to a degree, the extension of existing woodlands.

d. the areas of 'Other' private woodland were sampled intensively enough to provide quite independently the desired precisions of ±5% on total area and ±15% for the predominant forest type. Consequently, those counties with a higher proportion of 'Other' woodland which was subject to sampling error, and a lower proportion of Forestry Commission and Dedicated and Approved woodland data which has no sampling error, tend to have relatively poor precisions.

e. the percentage distribution of forest types within a county obviously has an effect. The woodland of the County of Kent, for example, contains 33 per cent Broadleaved High Forest and 32 per cent of Coppice. As there was no predominant forest type any increase in sample size would have had to be substantial to improve the precision of 13.2 per cent for Broadleaved High Forest and 12.9 per cent for Coppice.

Timber volume estimates

As no field work had been carried out in Forestry Commission or Dedicated and Approved woodlands a relatively straightforward method of volume computation had to be adopted using information already in our possession. In the case of Forestry Commission woodlands, use was made of the fact that in the database every component of every sub-compartment classed as High Forest has a planting year, a Yield Class and a net area recorded. By accessing the Management Tables, which were stored in the computer, it was possible to attribute a current notional standing volume to each individual component of the database. The volume ascribed to each component was reduced by 15 per cent to allow for gaps and rides. This volume could be allocated subsequently to diameter classes based on the assumed mean diameter of each stand.

In the case of Dedicated and Approved woodlands there was rather less information available but there was still quite a lot known. First, there was information on Yield Class by species and Conservancy for all private woodlands from the 1965 Census and these values were compared with those obtained for corresponding crops in 'Other' private woodlands in the 1980 Census. In general very few amendments needed to be considered and the 1965 Yield Class pattern was finally accepted virtually unchanged. When considering age there were very few cases where actual P Year had been recorded and consequently in its absence age was assumed to be the mean of the P Year class, e.g. crops planted between 1961 and 1970 were assumed to have a mean planting year of 1965, those planted between 1861 and 1900 were given a P Year of 1880, and Pre-1861 of P1851. Of course, planting during a decade is seldom evenly balanced and it is likely for example that P45 was too old a mean for the P41–50 class and P75 probably too young a mean for the P71–80 class. However, it was not thought advisable to try and adjust for this as the weighting might well have differed from Conservancy to Conservancy. Nevertheless, this generalised approach enabled a statement to be prepared for each county which recorded an area and a Yield Class for each species in each P Year Group. The volume calculation procedure was then similar to that used for Forestry Commission woodlands with the same allowances for the differences between gross and net areas being made.

For the estimation of timber volume in 'Other' private woodlands the volume of each species/age component in each wood was calculated and regression analysis similar to that adopted for the area calculations used to arrive at the estimates for each county. The ground survey technique involved the identification of relatively distinct and, if possible, homogeneous stands within each wood. One of three methods was then employed in order to estimate the volume of each sub-block.

1. Plots

Three or more circular plots of fixed size had been randomly sited within each sub-block and the diameter at breast height of every tree measured together with the height of the largest tree of each species. The basal area of each species in the sub block was first estimated using the following method:

$$G = \frac{\pi}{10\,000\,A} \sum_{i=1}^{i=n} \frac{d_i^2}{4}$$

where n = the number of trees in the plot
       G = basal area per hectare (m²/ha)
       A = total area of plots
      d_i = i th diameter measurement (cm)

Volume was then calculated using basal area and the main crop form height factors used to produce the tables in the *Forest mensuration handbook*[1]. These took the following forms:

$V = (C_1 + C_2H_t)G$ – for most conifers
$V = (C_1 + C_2H_t + C_3H_t^2)G$ – for Lawson cypress and all broadleaves except sycamore, ash and birch
$V = (C_1 + C_2 \ln H_t + C_3(\ln H_t)^2)G$ – for sycamore, ash and birch

where V = volume per hectare (m³/ha)
      H_t = mean top height (m)
      ln = natural logarithm
      C_1 = an estimated regression constant
   C_2, C_3 = estimated regression coefficients

A breakdown of volume into diameter classes was achieved by allocating the measured stems to their respective classes before applying the above formulae.

2. *Relascope sweeps*

The use of relascopes has already been discussed in Chapter 7. They were practical instruments in open coniferous stands, since basal area is recorded directly, and thus dispensed with the need for dbh measurements in each plot. Volume was calculated using the same formulae as above. A breakdown of volume into diameter classes was achieved by the use of stock tables[1].

3. *Thicket crops*

A third technique had been developed for thicket stage crops that were difficult to enter. Heights and diameters were obtained from trees on the stand edge and volumes calculated using mean height. Simple height/volume relationships were empirically derived for this purpose and took the form of a simple table for conifers and the following function for broadleaves.

$$V = 1.4\,(H_m - 6.0)$$

where V = volume per hectare (m³/ha)
      H_m = mean height

Whichever of the above three techniques was used volumes were only expressed for those species that formed a significant percentage of the sub-block canopy. These were referred to as component species. If a species of tree occurred in a plot but was not sufficiently abundant to be expressed as a component its volume was calculated and attributed to the species which formed the greatest proportion of the canopy, i.e. conifers being allocated to the coniferous component and broadleaves to the broadleaved component wherever possible. Sometimes the heights of these occasional trees were not recorded in which case they were allocated the mean top height of the species to which their volume was to be attributed.

Occasionally stands contained intimate mixtures of species and were simply described in whole or in part as mixed. For these stands all trees of a species other than one of the identifiable components were grouped together and their volume calculated using the appropriate basal areas and mean top heights. Those species without height measurements were allocated a top height from the species which belonged to the mixed component which had the most top heights recorded for it. A P Year class was always recorded for each of the main species but never recorded for individual stems. This created problems for crops of the same species which had been described as having two age classes. A method was therefore devised to divide the diameters measured during volume assessment into classes by separating them on either side of the median diameter. Provided at least one top height measurement was made for a tree falling either side of the median diameter then the volumes calculated from the larger girth trees were allocated to the older age class and the smaller to the younger. If top heights were not recorded for each diameter class then all volume was attributed to the older age class. For reasons of computational complexity no attempt was made to split volumes where three age classes were recorded for one species in a stand, or for two age classes where both age classes were classed as mixtures.

---

[1] Hamilton, GJ (1975). *Forest mensuration handbook*. Forestry Commission Booklet 39. HMSO.

CHAPTER 10

Calculation and presentation of the results – non-woodland trees

The volume estimates for the non-woodland tree survey were calculated in two stages:
  a. from aerial photographs; and
  b. from ground survey.
The procedures will be considered in that order.
The following data were obtainable from maps or aerial photographs:
  number of isolated trees
  number of clumps
  crown diameter of isolated trees
  area of clumps
  length of linear features
  area of linear features.

The first two were simple counts from the photographs, whilst the remainder were derived by direct measurements from maps or photographs.

The survey design consisted of a number of clusters of samples marked on aerial photographs which were in turn sub-sampled for ground survey. It was thus possible to confirm the correlation between each of the variables common to air and ground data (see Chapter 7). This was done for each sample stratum in each county. A program was written to retrieve the variables from the data files and plot scatter diagrams of the totals from each corresponding air and ground sample. The pairs of points plotted for each cluster were linked with lines to enable the 'within' cluster regression to be judged as well as 'between' cluster regression. Each of these scatter diagrams was subjectively judged on the basis of a number of criteria and a decision taken as to whether the aerial counts were sufficiently related to the ground data to justify incorporating all the air samples in the analysis to derive a final estimate. The decision was recorded in the computer file which was subsequently picked up by the analysis programs and the appropriate estimator applied.

The criteria by which the relationship between air and ground was judged were as follows:
  a. both 'within' and 'between' regression gradients should have been close to unity. If the slopes were too shallow or too steep or simply inconsistent it was indicative of some source of error or bias in the air photograph interpretation and hence should be treated with suspicion;

b. it was necessary to ensure that one point did not adversely dominate the regression. On occasions samples would arise with a high density of trees which was not reflected on the photograph or vice versa. Such spurious points could disproportionately affect the value of the regression coefficients;
c. it was also necessary to ensure that there were sufficient points to be confident of the relationships. In upland areas it was quite common for trees to be absent from all but one or two samples in which case there was serious doubt as to whether a regression relationship could be deemed representative. In these cases use of the air photograph data was rejected even if a regression coefficient of one was achieved. Of course if no trees occurred in the ground sample no relationship was capable of being derived;
d. correlation coefficients were calculated but no clear division was arrived at for an 'acceptable' correlation coefficient. Assessment of the degree of scatter or variation that was judged as being acceptable was essentially arrived at subjectively;
e. on occasions a large difference would arise between the tree counts for the air samples that were ground surveyed and those air samples that were not. This was suggestive of a large degree of variation in tree density that was not being captured by ground survey and would result in big differences between the estimates derived from ground only and air plus ground. This fact alone was not sufficient in itself to justify rejecting the air photograph data but necessitated a more critical look with regard to the above four criteria to safeguard against spurious estimates.
At the start of the Census the possibility of combining strata to form a relationship was considered if there was no evidence to suggest that the regression coefficient differed significantly between the two strata. This would have helped to overcome problems of data scarcity. It was rejected, however, when it became apparent that the combination of strata invariably resulted in a reduction of precision in the final estimates.

When the decision had been taken for each of the five factors for each sampling stratum they were recorded in a data file for subsequent retrieval by the analysis program. When the program was run it would calculate and estimate from ground data or ground plus air for each stratum as appropriate. The estimate for the county was simply the sum of the stratum totals.

Data collected on the ground survey consisted essentially of numbers of trees identified by species and health class with diameter and height recorded for each. This enabled estimates to be calculated for numbers and volumes of trees of each of the main species broken down into categories such as health or diameter classes. This was achieved simply by obtaining totals for each sample unit for each species/category and employing the two stage random sampling estimation routine for each stratum.

It was possible in some cases to bring the estimates from air and ground to bear on the estimates from ground only. The estimates of numbers of isolated trees of each species were a sub-category of the estimate of total numbers of isolated trees. Since the overall totals could be estimated more precisely using a combination of air and ground data it was appropriate to use them as the basis for adjusting the estimates of numbers of each species. This was simply achieved by multiplying each estimate by the ratio of (air and ground)/ground estimate. This ratio was normally close to unity. The estimates of volume of each species of isolated tree could be similarly adjusted but not estimates of numbers or volumes of trees in clumps and linear features since trees in these latter categories are not easily countable on aerial photographs.

The volume of each tree was calculated and accumulated to give a total volume for the sample unit. Using the ground survey data the stem volume (v) was calculated using species specific tariff number functions as follows:

$$v = 0.0052 - 0.0012t + 0.315tb - 0.0437b$$

where b = tree basal area (m^2)
$\quad\quad\quad t$ = tariff number

The tariff number was calculated using species specific coefficients ($C_1 - C_4$) as follows:

$$\text{Conifers } t = C_1 + C_2h + C_3d$$

where h = total height (m)
$\quad\quad\quad d$ = diameter at breast height (cm)

These functions were based on normally sized and shaped trees and hence an alternative formula $v = 0.3hb$ was used where

(i) $d < 10$ or $d > 80$
(ii) $h < 10$ or $h > 40$
(iii) $t < 10$

A fourth check was incorporated to ensure that the height of each tree was above the minimum for a given diameter employed in deriving these functions from the original sample.

$$\text{Broadleaves } t = C_1 + C_2h_m + C_3d + C_4\left(\frac{1}{d}\right)^2$$

where h_m = timber height (m)
$\quad\quad\quad b$ = tree basal area (m^2)

These functions again apply to normally sized and shaped trees. Alternative formulae were adopted if the diameters and/or heights fell outside certain specific limits, or if the derived tariff number fell below 10.

(i) $v = 0.98\ h_mb$ timber height too short
(ii) $v = 0.7\ h_mb$ acceptable height, diameter too small or too large
(iii) $v = 0.5\ h_mb$ timber height too tall
(iv) $v = 0.9\ h_mb$ tariff number <10

The method of volume calculation differed between isolated trees and clumps and linear features and can be summarised as follows:

a. if there was reasonable correspondence between the numbers of isolated trees found on the ground sample squares with those found on the same squares in the air count then the crown diameter/volume functions were generally used to calculate the total volume of the sample. This volume was then modified using information from the ground sample to make allowance for trees that had no volume because they were below 7 cm diameter at breast height, of poor stem form, or damaged or decayed; these are features that cannot be identified from aerial photographs. The adjusted volume was then subdivided by species, size classes, etc., on the basis of the ground sample, added to other data for the same stratum and the values raised to the stratum totals in proportion;

b. if there was poor correlation between the isolated tree counts and those obtained from the air count then the air count data were disregarded and the data worked up solely on the basis of the ground sample information;

c. in the case of clumps and linear features, where the correlation of crown diameter on volume was generally poor, the volume of the trees, and their associated details such as species, diameter classes, etc., were established for each clump or linear feature on the ground squares.

These details were then converted to a per unit area basis for the feature concerned, added to other data from the same stratum and applied to the estimated stratum totals derived from measurements made on the aerial photographs.

Monitoring precision targets

In the case of the non-woodland survey the monitoring program produced a summary showing for each cluster in the survey its cluster number, soil group and the values for a particular parameter in each ground visit and aerial survey square. The summary was supported by a listing of the mean and standard error for each sample stratum, for the whole county, and for both air and ground visit data. The parameters checked in this fashion were:

the number of isolated live trees
the number of all isolated trees both live and dead
the number of clumps
the area of clumps
the length of linear features
the area of linear features
the volume of live isolated trees
the area of 'Extra' woodland

Again, where possible, the monitoring was continuous once 50 per cent of the samples within a county had been completed.

Precision achieved

The range of sampling precisions on the number of isolated trees lies between 9.3 per cent in West Sussex and 30.3 per cent in Humberside. In the estimation of the numbers of all measurable trees in a county the highest precision is 8 per cent in Berkshire and the lowest is 33 per cent in Northumberland.

The non-woodland tree population proved to be very variable in some counties depending upon:

a. the effects of garden and urban trees, with high numbers, on the results of soil groups such as the fens with otherwise low tree populations;

b. the relative proportions of the county's area in such variable strata;

c. the proportion of trees in each of the main features of isolated trees, clumps and linear features.

Wherever possible extra samples were taken, as in Northumberland with a total of 39 clusters, but the cost of substantially increasing the sample was sometimes considered too high in relation to the numbers of trees and volume at issue.

Part II

RESULTS

CHAPTER 11

Total woodland area

The full results of the Census have already been published in the form of County, Conservancy and Country Reports and the discussion in this Chapter is therefore largely confined to an examination of the woodland area of Great Britain in relation to other countries in Europe, and the distribution of woodland between the three countries which comprise Great Britain.

Total woodland area

The total woodland area in Great Britain at 31 March 1980 was estimated to be 2 108 397 hectares. Of this total England held 947 688 hectares or 45 per cent, Wales 240 784 hectares or 11 per cent, and Scotland 919 925 hectares or 44 per cent. These figures are shown in Table 1 below together with the percentage that the woodland area represents of the total land and inland water area of each country.

Table 1 Total woodland area and woodland density as at 31 March 1980

Country	Woodland area (hectares)[1]	Per cent of total	Woodland density per cent of land and inland water area
England	947 688 ± 3 133	45	7.3
Wales	240 784 ± 803	11	11.6
Scotland	919 925 ± 2 837	44	12.6[2]
Great Britain	2 108 397 ± 4 303	100	9.4[2]

1. Although the Standard Errors of the estimates have been quoted in this instance they are not given in subsequent tables in this Report. The reader is referred to the Great Britain and the individual country, Conservancy and county Reports for this information.
2. Based on the land and inland water area of Scotland excluding the Western Isles, Orkneys and Shetlands. The Great Britain figure likewise excludes this area.

Comparison with total land and inland water area

From this Table it will be seen that in 1980 Great Britain had a percentage woodland cover of 9.4 per cent with the country values ranging from 7.3 per cent in England to 12.6 per cent in Scotland. However, the Scottish figure is based on the land area of Scotland exluding the Orkney, Shetland and Western Isles in which there is very little woodland. If the Scottish woodland area is expressed as a percentage of the total land area of Scotland then the figure is reduced to 11.7 and that for Great Britain is reduced to 9.2. Woodland density is sometimes expressed as a percentage of the land area of a country as distinct from the land and inland water area of the country and in this case the figure for England remains unaltered, Wales rises from 11.6 to 11.7 per cent and Scotland from 11.7 to 11.9 per cent.

Certain counties in England and Wales and Regions in Scotland have always had high densities of woodland but the afforestation programmes of recent years carried out both by private owners and the Forestry Commission have increased the densities quite markedly in some areas. The density of woodland for individual counties and Regions are listed in Appendix 4 and shown diagrammatically in Figure 2 but those with woodland densities of 12 per cent or more are shown below:

England	%	Scotland	%	Wales	%
Surrey	19	Dumfries and		West	
West Sussex	17	Galloway	21	Glamorgan	20
Hampshire	16	Strathclyde	15	Mid	
East Sussex	16	Grampian	15	Glamorgan	15
Northumberland	15	Borders	14	Powys	13
Berkshire	13	Central	13	Gwent	12

Comparison with other European countries

Table 2 compares the position of Great Britain in terms of total woodland area in relation to land area and per head of population with those of selected countries in Europe and with the European Economic Community.

Table 2 Woodland area in relation to total land area and population for selected countries in Europe

Country	Total woodland area (million hectares)	Percentage of land area	Area in hectares per head of population
Great Britain	2.1	9	0.04
Northern Ireland	0.1	5	0.04
United Kingdom	2.2	9	0.04
Belgium	0.7	23	0.07
Luxembourg	0.1	31	0.23
Denmark	0.5	11	0.09
France	15.1	28	0.28
Federal Republic of Germany	7.2	30	0.12
Greece	5.8	45	0.59
Netherlands	0.4	10	0.02
Ireland	0.4	6	0.11
Italy	8.1	27	0.14
EEC	40.3	25	0.15
Austria	3.8	46	0.50
Finland	23.2	76	4.82
Norway	8.7	28	2.12
Portugal	3.0	35	0.30
Spain	12.5	25	0.33
Sweden	27.8	68	3.34
Switzerland	1.1	28	0.18

Source: The forest resources of the ECE regions. FAO, 1985

Note: Figures may not total exactly due to rounding.

From this Table it will be seen that when woodland density is considered only the Republic of Ireland and Northern Ireland have a lower percentage than Britain whilst the Netherlands and Denmark have percentages which are the same as, or a little above, that of Great Britain. When woodland area in relation to population is considered it will be seen that of the selected countries only the Netherlands has a lower figure and the EEC average is some three times greater than that of Great Britain.

Ownership

Distribution of area between Forestry Commission and privately owned woodlands is shown in Table 3. In this Report, as in all previous Census Reports, the term private woodland refers to all woodland not owned or managed by the Forestry Commission. It includes woodland belonging to other Government Departments, nationalised industries and other statutory undertakings, local authorities and other public bodies although these agencies together probably account for less than 10 per cent of the private woodland total.

This table shows that in Great Britain 1 216 682 hectares, or 58 per cent of the total, are privately owned and the balance is held by the Forestry Commission. In England nearly three-quarters of the woodland total is in private hands whereas in both Wales and Scotland more than half is owned by the Forestry Commission. If each ownership class is considered separately then Scotland holds 56 per cent of the Forestry Commission holding, England 29 per cent and Wales 15 per cent. In the case of private woodlands England holds the bulk of the total 57 per cent, Scotland has 35 per cent and Wales 8 per cent.

Table 3 Total woodland area by ownership categories

Hectares

Country	FC woodlands		Private woodlands		Total woodland	
	Area	% of total	Area	% of total	Area	% of total
England	254 968	27	692 720	73	947 688	100
Wales	138 854	58	101 930	42	240 784	100
Scotland	497 893	54	422 032	46	919 925	100
Great Britain	891 715	42	1 216 682	58	2 108 397	100

Table 4 Area of woodland by forest type and county

All woodland ownerships   Hectares

Forest type	England		Wales		Scotland		Great Britain	
	Area	%	Area	%	Area	%	Area	%
Mainly coniferous high forest	382 497	41	167 960	70	766 351	83	1 316 808	62
Mainly broadleaved high forest	429 248	45	59 321	25	75 849	8	564 418	27
Total high forest	811 745	86	227 281	95	842 200	91	1 881 226	89
Coppice with standards	11 473	1	80	<1	15	<1	11 568	1
Coppice	25 711	3	1 849	1	4	<1	27 564	1
Total coppice	37 184	4	1 929	1	19	<1	39 132	2
Scrub	79 498	8	8 222	3	60 512	7	148 232	7
Cleared	19 261	2	3 352	1	17 194	2	39 807	2
Total	947 688	100	240 784	100	919 925	100	2 108 397	100

Forest type

Table 4 shows the distribution of the 2 108 397 hectares of woodland in Great Britain by countries and by forest type.

The first major point of note is the relatively high proportion of the area that is now classed as High Forest forming no less than 89 per cent of the Great Britain total. The values range from 95 per cent in Wales to 86 per cent in England. In both Wales and Scotland part of the reason is the high proportion of coniferous crops but nonetheless the substantial broadleaved area that can now be considered as being of suitable stocking and form to qualify for this category indicates the improvement in quality which has taken place since the war, and in particular since 1965. The natural process of selection, sometimes aided by intervention by man, and the fact that crops are older and more accessible has allowed better judgements to be made as to their current status and future potential than was the case in the two previous Surveys.

Coniferous crops predominate in the High Forest category in Great Britain in the ratio of about 7:3 but the three countries show substantial differences. In England there is slightly less Coniferous High Forest than Broadleaved High Forest, in Wales Coniferous High Forest accounts for about three-quarters of the High Forest total and in Scotland over 90 per cent of it.

Coppice types account for only 39 132 hectares or 2 per cent of the total woodland area. Simple Coppice occupies 27 564 hectares or 70 per cent of the coppice category. Virtually the whole area of Coppice with Standards is in England and also most of the Coppice, with the major part of the balance in Wales; Scotland holds negligible areas of either Coppice or Coppice with Standards.

Scrub and Cleared woodland account for 9 per cent of the Great Britain total, a percentage which is similar to that in both England and Scotland. Wales with only 4 per cent of its total area classed as unproductive has a substantially lower percentage than the others.

Planting year class

The breakdown of the High Forest area by planting year classes is shown in Table 5. Before discussing the implications of the results a number of points need to be made.

The first is that generally speaking the younger the crop is the easier it is to estimate its age if information on the actual year of planting is unavailable. This is one reason for the rigid 10 year tabular distribution back to 1901 and then only two age classes to cater for any crops planted prior to 1901.

The second point is that coniferous crops tend to be even aged and are consequently more likely to have a known specific planting year than broadleaved crops which may have arisen naturally and consequently may have a wide range of ages of trees present in the one crop. Also, the introduction of maiden stems into an existing crop, selective felling, or occasional cutting of coppice can likewise pro-

Table 5 Area of high forest types by planting year classes

All woodland ownerships

Hectares

Country	High forest type	Planting year classes																				
		P71–80		P61–70		P51–60		P41–50		P31–40		P21–30		P11–20		P01–10		P1861–1900		Pre 1861		Total	
		Area	%	Area	%	Area	%	Area	%	Area	%	Area	%	Area	%	Area	%	Area	%	Area	%	Area	%
England	Mainly coniferous	53 676	14	104 917	27	97 393	25	45 514	12	35 549	9	28 752	8	5 394	1	3 617	1	6 302	2	1 383	1	382 497	100
	Mainly broadleaved	15 745	4	28 953	7	54 030	13	52 562	12	38 500	9	40 840	9	26 244	6	33 716	8	93 254	22	45 404	10	429 248	100
	Total	69 421	9	133 870	16	151 423	18	98 076	12	74 049	9	69 592	9	31 638	4	37 333	5	99 556	12	46 787	6	811 745	100
Wales	Mainly coniferous	27 385	16	48 450	29	52 315	31	18 605	11	14 587	9	5 141	3	457	<1	436	<1	544	<1	40	<1	167 960	100
	Mainly broadleaved	705	1	1 318	2	6 451	11	14 112	24	4 877	8	4 012	7	7 311	12	5 630	10	10 125	17	4 780	8	59 321	100
	Total	28 090	12	49 768	22	58 766	26	32 717	14	19 464	9	9 153	4	7 768	3	6 066	3	10 669	5	4 820	2	227 281	100
Scotland	Mainly coniferous	247 569	32	212 439	27	160 563	21	52 666	7	37 445	5	25 251	3	6 346	1	6 862	1	11 887	1	5 323	1	766 351	100
	Mainly broadleaved	2 794	4	5 164	7	5 875	8	8 715	11	6 047	8	6 320	8	2 900	4	6 439	4	19 287	25	12 308	16	75 849	100
	Total	250 363	30	217 603	25	166 438	20	61 381	7	43 492	5	31 571	4	9 246	1	13 301	1	31 174	2	17 631	2	842 200	100
Great Britain	Mainly coniferous	328 630	25	365 806	28	310 271	24	116 785	9	87 581	7	59 144	4	12 197	1	10 915	1	18 733	1	6 746	<1	1 316 808	100
	Mainly broadleaved	19 244	3	35 435	6	66 356	12	75 389	14	49 424	9	51 172	9	36 455	9	45 785	6	122 666	22	62 492	11	564 418	100
	Total	347 874	18	401 241	21	376 627	20	192 174	10	137 005	7	110 316	6	48 652	3	56 700	3	141 399	8	69 238	4	1 881 226	100

duce a crop with a wide variety of ages. In these instances the crop was allocated to the P Year Class that seemed to best accommodate the bulk of the trees. Although the term Planting Year is used it also embraces crops which have arisen naturally through seeding or stump regeneration.

Thirdly, estimating the age of older crops does require skill and a knowledge of the rates of growth of species in various parts of the country and surveyors often needed to be reminded that broadleaved crops on the Welsh Marches or in southern England were much younger than they appeared when compared with crops of similar appearance in the north.

In the case of crops planted prior to 1901, where there was usually a lack of recorded information, there was a tendency for surveyors to allocate crops to the 1861–1900 P Year class rather than the pre-1861 class on the grounds that the trees were more likely to have a mean age of a hundred years rather than the implied mean of about 140–150 years of the older class. The areas of the crops in the older P Year classes, and in particular the two oldest classes, must therefore be considered as being best estimates in that they have been allocated in many instances to a P Year class on the basis of assumed age rather than on firm evidence.

If each of the forest types is dealt with in turn the most striking aspect is the very large proportion of the coniferous area which has been planted since the war. No less than 85 per cent of the Mainly Coniferous area has been planted since 1940. The bulk of the increase in the two decades after the war came from the major afforestation schemes of the Forestry Commission with private owners tending to concentrate on the replanting of crops felled either during the war, or in the immediate post-war period when the pressure of felling was still strong. However, private afforestation on an increasing scale began in the sixties and has continued, apart from a major drop in the mid and late seventies, to the present day. Indeed, the area of new planting by private owners now exceeds that of the Forestry Commission. The areas planted before the last war show a fairly predictable pattern in that crops established by the Forestry Commission and private owners were generally too young to produce much in the way of utilisable material and consequently largely escaped the war-time fellings. There is, therefore, still a relatively large area in crops planted between 1921 and 1940 but these crops are now approaching maturity and it can be expected that the areas in these classes will diminish over the next 15 years as crops are felled. Crops planted prior to the First World War are now relatively uncommon. Many of them comprise woodlands in the grounds of the larger estates where they are retained for shelter, amenity and sporting purposes.

The pattern in the three countries is similar in that they all exhibit a concentration in the post-1941 era. However, England has about three-quarters of its Mainly Coniferous area in this category while Wales and Scotland, where the bulk of the post-war afforestation took place, have 87 per cent. The position in the inter-war period is reversed with England holding 17 per cent of this category, Wales 12 per cent and Scotland 8 per cent. In the oldest P Year classes there is now only a very limited area of old coniferous crops in Wales whilst Scotland and England hold about 5 per cent of their totals in these classes.

The broadleaved pattern in Great Britain is markedly different from that of conifers. Only 35 per cent of the area has been planted since the war, 18 per cent in the inter-war years and 47 per cent prior to 1920. There is thus a much more balanced age class distribution although a more detailed examination shows peaks and troughs within the broad age bands.

Two features are immediately obvious. The first is the relatively large area still contained in crops planted prior to 1900. However, with broadleaved rotations generally at least 100 years, and usually considerably longer, one would expect about a third of the total to be in crops of 80 years and over and in fact this is the position. The second point of note is the peaking of the P Year classes during the period 1941–60. This, however, is understandable if it is remembered that many of these crops are ones which arose from fellings during the Second World War, either from broadleaved crops cleared during the period or as a result of broadleaved species naturally regenerating sites which had previously carried coniferous or broadleaved crops. In many instances these sites were left undisturbed for a number of years after felling and by the time arrangements had been made to bring them back into production there was an almost impenetrable thicket which would have been very costly to clear. Consequently the crops were often allowed to develop naturally with, in some cases, cleaning and thinning carried out at a later stage to encourage the best stems. It is likely that many of the crops on sites felled in the First World War and in the twenties and thirties also arose in this fashion.

This pattern was not so clearly evident in the 1965 Census as many crops which had arisen from Second World War fellings were under 20 years of age, of variable form and stocking, not yet showing their potential in terms of form and diameter growth, and so were classed as Utilisable, and even in some cases Unutilisable, Scrub. What is apparent, however, is that after this major increase in broadleaved area arising mainly by natural means the amount of

Table 6 Area of high forest by principal species and countries

All woodland ownerships

Hectares

Species	England Area	England % of category	England % of all species	Wales Area	Wales % of category	Wales % of all species	Scotland Area	Scotland % of category	Scotland % of all species	Great Britain Area	Great Britain % of category	Great Britain % of all species
Scots pine	91 074	24	11	5 592	3	2	144 371	19	17	241 037	18	13
Corsican pine	40 212	10	5	3 693	2	2	3 346	—	—	47 251	4	2
Lodgepole pine	15 249	4	2	7 895	5	3	103 924	13	12	127 068	10	7
Sitka spruce	75 599	20	9	85 701	51	38	364 601	48	44	525 901	40	28
Norway spruce	43 499	11	5	18 641	11	8	54 707	7	7	116 847	9	6
European larch	21 862	6	3	2 595	2	1	15 957	2	2	40 414	3	2
Jap./Hybrid larch	35 742	9	5	23 461	14	10	52 146	7	6	111 349	8	6
Douglas fir	25 063	6	3	10 708	6	5	11 628	2	1	47 399	4	2
Other conifers	17 442	4	2	7 906	5	4	6 312	1	1	31 660	2	2
Mixed conifers	21 665	6	3	1 744	1	1	8 641	1	1	32 050	2	2
Total conifers	387 407	100	48	167 936	100	74	765 633	100	91	1 320 976	100	70
Oak	129 352	30	16	26 087	44	11	16 551	22	2	171 990	31	9
Beech	57 828	14	7	5 612	10	2	10 496	14	1	73 936	13	4
Sycamore	36 204	9	4	3 833	6	2	9 389	12	1	49 426	9	3
Ash	56 092	13	7	9 387	16	4	4 102	5	1	69 581	12	4
Birch	45 901	11	6	5 583	9	2	16 647	22	2	68 131	12	4
Poplar	12 757	3	2	504	1	<1	329	—	<1	13 590	2	1
Sweet chestnut	9 451	2	1	412	1	<1	8	—	<1	9 871	2	<1
Elm	5 545	1	1	358	1	<1	3 611	5	<1	9 514	2	<1
Other broadleaves	20 294	5	2	3 718	6	2	5 115	7	1	29 127	5	2
Mixed broadleaves	50 914	12	6	3 851	6	2	10 319	13	1	65 084	12	3
Total broadleaves	424 338	100	52	59 345	100	26	76 567	100	9	560 250	100	30
Total	811 745	100	100	227 281	100	100	842 200	100	100	1 881 226	100	100

Note: The total area of conifer and of broadleaved species in this and subsequent High Forest tables differs from the total area classified as Mainly conifer and Mainly broadleaved in Tables 4 and 5. Here areas have been allocated on the basis of species distribution and not forest type distribution.

broadleaved planting and regeneration has declined markedly and there has been relatively limited planting or regeneration of broadleaves in recent years.

However, the future position is almost certainly not as bleak as the figures show. First, the very evident concern of conservation bodies and the public in general about the future of broadleaved trees has resulted in substantial increases in planting since 1980, the operative date of the Census. Secondly, where broadleaved species have been planted in mixture with conifers such crops were assigned to Mainly Coniferous or to Mainly Broadleaved High Forest on the basis of current proportions. Thus a crop which is predominantly coniferous at the moment, although grown with the intention that it should eventually become a pure or mainly broadleaved crop in time, will have had the total area attributed to Mainly Coniferous High Forest. The areas present in the three youngest P Year classes can therefore be expected to increase in future years as the 50 per cent threshold between Mainly Coniferous and Mainly Broadleaved is reached and passed.

The planting year patterns within the three countries follow the same trend as that for Great Britain. England has 32 per cent of the total crops of pre-1900, Wales 25 per cent and Scotland 41 per cent but these, and subsequent percentages need to be looked at in the light of the fact that the bulk of the broadleaved resource occurs in England and variations in the distribution in the other two countries have a relatively minor effect on the overall position. The percentages in the inter-war P Year classes are similar in all three countries while in the P41–60 age group England has 25 per cent of the area in these two classes, Wales 35 per cent and Scotland 19 per cent. The drop in area in the planting classes younger than P60 is more marked in Wales than in England or Scotland. The age class distribution, whilst interesting in itself does, however, also need to be looked at in terms of the species distribution because if crops are under active management they will have different rotations and if unmanaged they will have different life spans. Clearly if a certain species dominates certain P Year classes this must have implications for the future.

Composition by principal species

The distribution of High Forest area by species for the three countries is shown in Table 6.

It is advisable at this stage to point out again the fact that there are slight differences in the coniferous and broadleaved areas quoted in Tables 5 and 6 although the High Forest totals are unaltered. In the case of Table 5, crops have been allocated to Mainly Coniferous and Mainly Broadleaved High Forest on the basis that crops which were more than 50 per cent composed of coniferous species by canopy, or in young crops by numbers of trees, were allocated to Mainly Coniferous High Forest and all others to Mainly Broadleaved. In the case of Table 6, and Table 7 which follows it, the coniferous element in a crop has been extracted and apportioned to the species involved irrespective of whether the crop was classed as Mainly Coniferous or Mainly Broadleaved; the broadleaved element has been treated likewise. The differences between the two approaches result in only minor differences in area.

In Great Britain coniferous species account for 70 per cent of the total High Forest area and the broadleaved species for the remaining 30 per cent. Sitka spruce is the most widespread species occupying more than half a million hectares and accounting for 28 per cent of the total High Forest area. Scots pine is the next most important species in area terms with 13 per cent followed by oak with 9 per cent and Lodgepole pine with 7 per cent. The pattern, however, differs from country to country. In England conifers account for slightly less than half the High Forest area and consequently oak is the dominant species followed by Scots pine and Sitka spruce, these three species accounting for 36 per cent of the total area. However, there are also substantial areas of beech, ash and birch in the broadleaves and Corsican pine, Norway spruce and Japanese/Hybrid larch in conifers. In Wales on the other hand Sitka spruce is by far the most widespread species with 38 per cent of the total followed by oak with 11 per cent. Japanese/Hybrid larch, Norway spruce and, to a much lesser extent, Douglas fir are the only other species with relatively substantial areas. Scotland with a heavy preponderance of coniferous crops has Sitka spruce with 43 per cent of the total High Forest area followed by Scots pine with 17 per cent, Lodgepole pine with 12 per cent and Norway spruce and Japanese/Hybrid larch both with 6 per cent. Oak and birch are the two dominant broadleaves with 2 per cent each. The distribution pattern for Great Britain is shown in Table 7.

Although Sitka spruce is reasonably represented in area terms from 1921 onwards it will be seen that there has been an increasing use of this species and that about 40 per cent of its total area has been planted in the last decade. The same general pattern is true of Norway spruce but the use of this species has declined quite markedly within the last decade.

The fluctuating fortunes of Scots pine are also evident. Quite a substantial area of older crops still remain, some being the remnants of the Old Caledonian Pine Forest in the pre-1861 age class but there is a relative shortage of crops planted between 1901

Hectares

Table 7 Area of high forest by principal species and planting year classes – Great Britain

All woodland ownerships

Species	P71–80	P61–70	P51–60	P41–50	P31–40	P21–30	P11–20	P01–10	P1861–1900	Pre-1861	Totals
Scots pine	15 378	48 538	67 609	27 638	22 865	27 605	6 244	6 417	13 306	5 437	241 037
Corsican pine	8 416	13 451	10 880	4 111	5 473	3 723	492	294	346	65	47 251
Lodgepole pine	48 797	50 240	25 558	1 022	1 085	346	18	2	—	—	127 068
Sitka spruce	208 855	153 888	90 470	42 583	22 558	6 835	367	110	184	51	525 901
Norway spruce	10 337	31 041	32 253	18 233	16 473	6 367	1 018	458	569	98	116 847
European larch	2 603	6 179	7 379	4 906	7 050	4 750	2 057	2 696	2 422	372	40 414
Jap./Hybrid larch	19 260	24 135	41 840	11 518	6 928	3 694	601	225	131	17	111 349
Douglas fir	6 497	15 805	14 216	2 827	2 534	4 156	586	244	335	199	47 399
Other conifers	4 886	13 697	9 276	1 130	991	651	168	185	530	146	31 660
Mixed conifers	4 360	8 495	7 586	2 537	2 015	2 081	1 273	972	2 109	622	32 050
Total conifers	329 389	365 469	310 067	116 505	87 972	60 208	12 824	11 603	19 932	7 007	1 320 976
Oak	2 228	2 863	7 317	11 692	9 423	12 133	12 497	20 143	60 965	32 729	171 990
Beech	2 510	6 313	11 057	4 577	3 968	2 500	2 725	5 232	18 098	16 956	73 936
Sycamore	2 383	4 375	5 920	9 004	6 319	4 882	3 837	3 650	7 187	1 869	49 426
Ash	1 722	2 706	7 615	13 524	9 854	11 151	5 826	6 475	8 747	1 961	69 581
Birch	2 776	8 820	17 194	20 341	8 759	5 367	2 758	1 103	852	161	68 131
Poplar	2 007	4 463	4 732	1 032	326	594	301	38	75	22	13 590
Sweet chestnut	354	366	1 143	1 147	1 163	978	769	973	1 480	1 498	9 871
Elm	156	219	356	1 045	948	1 360	651	1 012	3 132	635	9 514
Other broadleaves	1 521	1 903	4 738	6 550	2 806	5 026	1 914	1 192	2 777	700	29 127
Mixed broadleaves	2 831	3 768	6 527	6 769	5 464	6 115	4 533	5 259	18 136	5 682	65 084
Total broadleaves	18 488	35 796	66 599	75 681	49 030	50 106	35 811	45 077	121 449	62 213	560 250
Total	347 877	401 265	376 666	192 186	137 002	110 314	48 635	56 680	141 381	69 220	1 881 226

and 1920 with many of their contemporaneous stands having been felled for pitwood and other uses during the last war. There still remain substantial areas of middle aged crops and the species was used extensively in woodland rehabilitation after the war. Since then, however, its use has declined. Generally, the bulk of the conifer crops are young and the only species with significant areas in the older age classes are Scots pine and European larch.

With broadleaves, however, the pattern is quite different. As already noted there is a generally good distribution over the age class range but this is not necessarily so when individual species are considered. Oak, for example, is still the predominant species but more than half its total area now occurs in age classes dating from before 1900. There are reasonable quantities in most of the age groups up to 1950 but thereafter there has been an apparent decline in its planting and the areas involved in the last two decades have been relatively small. This may be due in part to the fact that it is the species which is often used in mixture and interplanted through existing crops and it is quite possible that the areas shown under-estimate those that will be predominantly oak in future. Also, it is a species which is slower growing at the outset than many other broadleaved species but in time can come to dominate the upper canopy as the other species are either overtopped or die out. Birch, on the other hand, is a relatively short-lived species and over 70 per cent of its area is under 40 years of age. This species was favoured by the growing conditions produced by war-time fellings and it is unlikely that the areas planted or regenerated in the future will ever again attain such levels; consequently the area of this species will probably decline as the crops age or other more economic species are introduced. Beech, for many years a major element of broadleaved planting, has also shown something of a decline. There still remain substantial areas in the older age groups, indeed about 50 per cent is classed as pre-1901, but the areas in the other age classes are relatively small. However, it was widely used in post-war restocking and it ranks second to birch in importance in broadleaved crops dating from the last three decades. Ash and sycamore show similar patterns in that both species have substantial areas in the older P Year groups and also appear to have colonised widely, often at the expense of oak, following the fellings of the First and Second World Wars. In terms of area sycamore in fact has done as well as ash over the last 30 years and currently seems to be establishing itself at a rather faster rate. Elm, never really significant as a woodland species in area terms where it usually occurs in mixture rather than pure, has only small areas in the younger age classes and it is likely that a substantial part of the 9500 hectares quoted has succumbed to Dutch elm disease since the Census was completed.

Table 8 shows the principal species in each of the planting year classes as at 1980. It should be remembered though that the table represents the pattern of crops as it exists *now* and because of the different felling ages does not necessarily reflect the relative importance of individual species in earlier years.

Table 8 Principal species in high forest by planting year classes – Great Britain

All woodland ownerships

Planting Year Classes	Principal species by percentage of area					
	First	%	Second	%	Third	%
P71–80	Sitka spruce	60	Lodgepole pine	14	Jap./Hybrid larch	6
P61–70	Sitka spruce	38	Lodgepole pine	13	Scots pine	12
P51–60	Sitka spruce	24	Scots pine	18	Jap./Hybrid larch	12
P41–50	Sitka spruce	22	Scots pine	14	Birch	11
P31–40	Scots pine	17	Sitka spruce	16	Norway spruce	12
P21–30	Scots pine	25	Oak	11	Ash	10
P11–20	Oak	26	Scots pine	13	Ash	12
P01–10	Oak	36	Ash	11	Scots pine	11
P1861–1900	Oak	43	Mixed broadleaves	13	Beech	13
Pre 1861	Oak	47	Beech	25	Mixed broadleaves	8

Oak currently is the most important individual tree species in all P Year classes prior to 1920. For the next two decades it is Scots pine and then, as a consequence of the very substantial afforestation programmes, Sitka spruce is dominant. The second most important species in age class terms shows again that the broadleaved species dominate the older classes, and that Scots pine is significant in the post-war planting but has now been superseded by Lodgepole pine. Japanese/Hybrid larch only attains third position in the P51–60 and P71–80 classes and European larch, once a major species on private estates, has dropped out of the list. This fact perhaps goes to emphasise that with so much of the post-war planting being afforestation it has meant a move away from those species which up until the advent of machinery capable of ploughing the upland peats and heathland had almost completely dominated the plantation scene. Their unsuitability for these new, much poorer and more exposed situations meant not only their decline in terms of the area planted but also that the existing crops as they come to maturity are likely to be replaced by faster growing species.

Coppice and coppice with standards

To be included in one or other of these two categories the critical factor was the condition of the coppice. It had to have at least two stems per stool and had either to be being worked on a known rotation or be composed of marketable species and so capable of being worked. The Census definition also laid down that at least half the stems had to have the capability of producing 3 m lengths or more which were of good form and the mean diameter of the crop had not to exceed 15 cm dbh. The necessary qualifications resulted in many coppice type crops being rejected either because they were not of a marketable species, were of poor form or were of such a diameter that they had to be considered as being Broadleaved High Forest of Coppice origin rather than as Coppice. Such overgrown crops can of course be reconverted to Coppice but at the time of inspection they were certainly well past the normal rotation age for cutting. In the case of Coppice with Standards the coppice had to meet the criteria listed above and in addition there had to be an overstorey of at least 25 stems per hectare which were at least one rotation older than the coppice. The results are shown in Tables 9 and 10.

Table 9 Area of coppice by principal species – Great Britain

All woodland ownerships   Hectares

Sub type		Principal species of coppice						Total
		Sycamore	Ash	Sweet chestnut	Hornbeam	Hazel	Other species	
With standards	ha	119	193	5 275	1 697	1 465	2 819	11 568
	%	1	2	45	15	13	24	100
Coppice only	ha	2 380	1 554	13 816	1 716	1 630	6 468	27 564
	%	9	6	50	6	6	23	100
Total	ha	2 499	1 747	19 091	3 413	3 095	9 287	39 132
% of coppice total		6	4	49	9	8	24	100

Note: 'Other species' includes mixtures of the above five named species as well as other minor species of coppice, e.g. birch and oak.

Table 10 Area of coppice with standards by principal species of both coppice and standards – Great Britain

All woodland ownerships   Hectares

Principal species of standard	Principal species of coppice						Total
	Sycamore	Ash	Sweet chestnut	Hornbeam	Hazel	Other species	
Conifers	—	—	16	4	—	—	20
Oak	101	173	4 897	1 594	1 444	2 792	11 001
Ash	8	20	—	88	21	27	164
Sweet chestnut	—	—	353	—	—	—	353
Other broadleaves	10	—	9	11	—	—	30
Total	119	193	5 275	1 697	1 465	2 819	11 568
% of total	1	2	45	15	13	24	100

In all, 39 132 hectares were recognised of which 11 568 hectares or 30 per cent were Coppice with Standards and 27 564 hectares or 70 per cent Coppice. The area of Coppice with Standards appears to have changed little in the last 15 years but there has been a 50 per cent increase in the coppice area. Part of this may be due simply to an increased working of young broadleaved crops, often for fuelwood, but part is undoubtedly due to a renewed interest in this form of management and a number of instances are known where overgrown Coppice or Coppice with Standards crops have been converted to Coppice either by cutting, layering, or actual planting with a view to working such crops on rotation in future. England holds virtually all the Coppice with Standards and most of the Coppice, with Wales holding only 7 per cent of the latter category.

When coppice species are considered, whether in Simple Coppice or underneath Standards, Sweet chestnut is the dominant species, accounting for about half the total, with significant quantities of hornbeam, hazel, sycamore and ash. The first two figure prominently in both Coppice and Coppice with Standards whereas sycamore and ash appear most frequently as Simple Coppice. It is quite possible that these last two named species may well have been intrusive at some stage and have been coppiced along with the other species in the stand. There is a substantial area of Coppice classed as 'Other species', some 24 per cent of the total, but this category not only includes individual species not listed elsewhere in the Tables but also caters for mixtures of named species. An examination of the species of standard in Coppice with Standards shows that it is almost invariably oak.

Scrub

The total area of Scrub in Great Britain was estimated to be 148 232 hectares or 7 per cent of the total forest area. Scrub was defined as inferior crops where more than half the trees were of poor form, poor timber potential or composed of unmarketable species and so did not qualify as either High Forest or Coppice. This definition therefore catered not only for crops experiencing poor site conditions where the form of the trees is also likely to be poor, but also lower timber quality crops occurring on sites which are not inherently poor and are consequently capable of growing something better. In addition, Scrub also included crops which, irrespective of site conditions, are composed of woody species which are currently unutilisable such as rhododendron, thorn, sallow, etc. The distribution of the Scrub area by countries and species is shown in Table 11.

Table 11 Area of scrub by principal species

All woodland ownerships   Hectares

Principal species	England		Wales		Scotland		Great Britain	
	Area	% of total	Area	% of total	Area	% of total	Area	% of total
Conifers	824	1	5	<1	1 615	3	2 444	2
Oak	12 568	16	1 382	17	4 104	7	18 054	12
Beech	775	1	38	<1	257	<1	1 070	1
Sycamore	1 911	2	82	1	373	1	2 366	2
Ash	7 276	9	224	3	376	1	7 876	5
Birch	20 390	26	1 128	14	41 742	68	63 260	42
Sweet chestnut	256	<1	—	—	8	<1	264	<1
Alder	4 736	6	1 375	17	2 252	4	8 363	6
Hornbeam	400	1	10	<1	—	—	410	<1
Hazel	5 171	6	965	12	2 425	4	8 561	6
Willow	4 005	5	367	4	592	1	4 964	3
Other broadleaves	21 186	27	2 646	32	6 768	11	30 600	21
Total	79 498	100	8 222	100	60 512	100	148 232	100

From this table it will be clear that England contained 79 498 hectares or 54 per cent, Wales 8222 hectares or 5 per cent and Scotland 60 512 hectares or 41 per cent. Birch is the predominant species in Scrub in Great Britain occupying 42 per cent, followed by oak with 12 per cent, and hazel, alder and ash with approximately 6 per cent each. Birch is also the most widely represented individual species in Scotland, where it occupied 68 per cent, and in England where it represented 26 per cent. But in Wales oak and alder with 17 per cent each predominate and birch takes only third place. Birch is a species which is often one of the first colonisers and can quickly form fairly dense thickets. As birch grows older, however, stocking often becomes poorer as age and windblow take their toll and if this situation is coupled with grazing pressure the trees in time become moribund and the area classifiable as woodland gradually declines.

Oak is very often semi-natural in origin and occurs widely throughout the country. Its classification as Scrub is sometimes due to the past treatment of the crop, in others to poor site factors. However, this species, in common with others such as birch, can often be upgraded at a later stage either through natural growth or through human intervention.

It was noticed during the course of the 1980 Census that crops which could be considered now as being of High Forest quality would more likely have been classed as Scrub a few years ago. This is not necessarily attributable to any change of definition but rather that the faster growing stems have by now begun to dominate the crop and there were enough of these trees of better form to justify upgrading it. Given adequate site conditions and reasonable shelter even the most unpromising crops of timber trees are, in time, capable of accession to one of the productive categories.

The areas of species such as Sweet chestnut, hornbeam and hazel are probably, in many cases, the remains of abandoned coppice crops.

Cleared

Cleared woodland is the category used to describe areas which had been felled and where the stumps were still in evidence or crops where the canopy stocking had been reduced to less than 20 per cent. The implication in both cases is that these areas will in time be replanted, naturally regenerated, or colonised so that they will remain as woodland. The total area involved was just under 40 000 hectares, or 2 per cent of the total, of which just under half lay in England. Scotland held the bulk of the remainder and Wales had only 8 per cent.

CHAPTER 12

Private woodland area

The breakdown of the 1 216 682 hectares of woodland in private ownership is shown in Table 12.

Forest type

From this table it will be seen that 83 per cent of the area is classed as High Forest with slightly more than half of this total composed of broadleaved species. Coppice and Coppice with Standards together account for just over 38 000 hectares with the area of Coppice more than double that of Coppice with Standards. Scrub occupies over 142 000 hectares or 12 per cent of the total and Cleared more than 28 000 hectares or 2 per cent of the total.

The pattern of distribution, however, differs between countries. In England, whilst the High Forest proportion is close to that of Great Britain there is more than twice as much broadleaved area as there is coniferous. England also holds nearly all the Coppice and Coppice with Standards and a little more than half the Scrub. Wales on the other hand has a higher proportion of its area in productive crops than the other two countries. Its High Forest area is predominantly broadleaved, such Coppice crops as exist are mainly Simple Coppice and the Scrub and Cleared totals combined account for only 8 per cent of the total. Scotland is characterised by a preponderance of Coniferous High Forest with Broadleaved High Forest accounting for only about 20 per cent of the High Forest category. Coppice types are virtually unrepresented but Scrub, at nearly 58 000 hectares, occupies a rather higher proportion of the overall total than is the case in the other two countries. A much lower proportion of Scrub recorded in the current Census when compared with those of the past is partly due to natural development of the crops and partly to clearance and replanting.

The characteristics of the various forest types in Total Woodlands have already been discussed in Chapter 11 and consequently comment here is largely confined to points of interest specific to private woodlands.

Table 12   Area of woodland by forest type and country

Private woodlands   Hectares

Forest type	England		Wales		Scotland		Great Britain	
	Area	%	Area	%	Area	%	Area	%
Mainly coniferous high forest	178 118	26	38 030	37	281 357	67	497 505	41
Mainly broadleaved high forest	385 272	56	53 198	52	71 816	17	510 286	42
Total high forest	563 390	82	91 228	89	353 173	84	1 007 791	83
Coppice with standards	11 441	2	80	<1	15	<1	11 536	1
Coppice	24 712	3	1 848	2	4	<1	26 564	2
	36 153	5	1 928	2	19	<1	38 100	3
Scrub	77 488	11	7 205	7	57 789	14	142 482	12
Cleared	15 689	2	1 569	2	11 051	2	28 309	2
Total	692 720	100	101 930	100	422 032	100	1 216 682	100

Planting year class

Table 13 analyses the Mainly Coniferous and Mainly Broadleaved High Forest totals by P Year class.

This Table shows more clearly than any other the differing age class patterns between conifers and broadleaves in private woodlands in Great Britain. For example about 75 per cent of the coniferous area has been planted since the war compared with only 20 per cent of the broadleaved area. All three countries show the heavy weighting to young coniferous crops but this feature is more pronounced in Wales than in the other two countries. No less than 85 per cent of the Welsh conifers are 30 years of age or younger while the area over 80 years of age is approximately one per cent. Scotland has 78 per cent planted since 1950 and 6 per cent planted prior to 1901 whilst corresponding percentages for England are 67 per cent and 4 per cent. In England and Wales the P61–70 age class is the one with the highest area and the area in the P71–80 age class is well down on that in the previous decade. In Scotland, however, the area planted in each decade has increased since the war and over a third of the coniferous area is now contained in the youngest age class.

The broadleaved pattern, on the other hand, shows a declining area planted in all three countries in each decade since the war. Each country also holds between 15 and 18 per cent of its area in the two decades between the wars and also has the bulk of the area in the pre-1901 classes. In England the proportion in crops of 80 years of age and over is 33 per cent, in Wales 25 per cent and in Scotland 43 per cent.

Composition by principal species

In Great Britain the High Forest category is split equally between conifers and broadleaves with oak the predominant species overall and Sitka spruce the dominant conifer (Table 14). The gap between oak and the next most important species Sitka spruce is, however, narrowing annually and with the High Forest area increasing largely as a result of upland afforestation it is unlikely to be long before Sitka spruce is the most widely planted species in private woodlands in the country.

In England one-third of the High Forest area is coniferous and two-thirds broadleaved. Scots pine is still the predominant conifer in private woodlands followed by Japanese/Hybrid larch and Norway spruce. European larch still has a significant proportion of the total but Sitka spruce, now fourth in order of importance, can be expected to increase in future at the expense of most other species as a consequence of its use in afforestation schemes and also for restocking. Of the broadleaved species oak is dominant, accounting for nearly a third of the total, and it is followed by ash, birch, beech and sycamore. The Mixed broadleaves category is third in importance and consists of a wide range of species which may include some that are noted immediately above.

In Wales, Sitka spruce is predominant among conifers and accounts for more than half the coniferous area. The only other species with substantial areas are Japanese/Hybrid larch, Norway spruce and Douglas fir with all four species combined accounting for over 80 per cent of the coniferous total. Nearly 60 per cent of the High Forest area is broadleaved with oak dominant and accounting for 45 per cent of the category; oak in fact has a greater area than any other species, coniferous or broadleaved, in High Forest; ash with 17 per cent is second in importance in the broadleaves followed by birch with 10 per cent. Sycamore and beech are the only other species important in area terms.

In Scotland, Sitka spruce is now clearly the most important species with 38 per cent of the coniferous area and is followed by Scots pine with 30 per cent. This change in relative importance has come about over a relatively short space of time as Scots pine accounted for more than half the Coniferous High Forest total as recently as 1965. The very substantial increase in afforestation over the latter period, however, has altered the pattern dramatically and the importance of Sitka spruce can be expected to increase further. Other species with fairly substantial areas are Norway spruce, Lodgepole pine and Japanese/Hybrid larch. Of these species only Lodgepole pine is likely to show any significant increase but with the use of pure Lodgepole pine in afforestation schemes now rather uncommon its rise in area is not likely to be as dramatic as might have been forecast a few years ago. Birch is the most important broadleaved species and is closely followed by oak. There are also, however, substantial areas of beech and sycamore with roughly equivalent areas of each and elm still occupies a significant proportion of the whole.

The P Year class distribution by species in Great Britain as a whole is shown in Table 15.

From this Table it will be seen that Sitka spruce owes its current importance to planting in the last 20 years and the more traditional private woodland species such as Scots pine, Norway spruce and European larch have all shown a substantial decline in area since the war. Pioneering species such as Japanese/Hybrid larch which were once used extensively in upland afforestation have likewise shown a drop in area, while the extent of the poor quality land now being planted is shown by the recent sizeable increase in the area of Lodgepole pine. At the older end of the age range Scots pine and European

Table 13 Area of high forest types by planting year classes

Private woodlands

Hectares

Country	High forest type	Planting year classes																				Total	
		P71-80		P61-70		P51-60		P41-50		P31-40		P21-30		P11-20		P01-10		P1861-1900		Pre 1861			
		Area	%	Area	%	Area	%	Area	%	Area	%	Area	%	Area	%	Area	%	Area	%	Area	%	Area	%
England	Mainly coniferous	26 857	15	56 446	31	37 145	21	17 453	10	14 230	8	11 279	6	4 667	3	3 119	2	5 697	3	1 225	1	178 118	100
	Mainly broadleaved	14 520	4	25 646	7	42 567	11	45 547	12	32 577	8	37 685	10	25 118	6	32 928	9	90 018	23	38 666	10	385 272	100
	Total	41 377	7	82 092	15	79 712	14	63 000	11	46 807	8	48 964	9	29 785	5	36 047	6	95 715	17	39 891	7	563 390	100
Wales	Mainly coniferous	9 146	24	15 235	40	7 964	21	1 770	5	1 321	3	1 350	4	363	1	397	1	449	1	35	<1	38 030	100
	Mainly broadleaved	580	1	933	2	4 071	8	13 462	26	4 345	8	3 707	7	6 928	13	5 557	10	9 121	17	4 494	8	53 198	100
	Total	9 726	11	16 168	18	12 035	13	15 232	17	5 666	6	5 057	6	7 291	8	5 954	6	9 570	10	4 529	5	91 228	100
Scotland	Mainly coniferous	93 584	34	79 050	28	45 500	16	13 941	5	9 696	3	11 789	4	5 736	2	6 488	2	10 588	4	4 985	2	281 357	100
	Mainly broadleaved	2 413	3	4 972	7	5 135	7	8 249	11	5 437	8	6 070	8	2 755	4	6 390	9	18 406	26	11 989	17	71 816	100
	Total	95 997	27	84 022	24	50 635	14	22 190	6	15 133	4	17 859	5	8 491	2	12 878	4	28 994	8	16 974	5	353 173	100
Great Britain	Mainly coniferous	129 587	26	150 731	30	90 609	18	33 164	7	25 247	5	24 418	5	10 766	2	10 004	2	16 734	2	6 245	1	497 505	100
	Mainly broadleaved	17 513	3	31 551	6	51 773	11	67 258	13	42 359	8	47 462	9	34 801	7	44 875	9	117 545	23	55 149	11	510 286	100
	Total	147 100	15	182 282	18	142 382	14	100 422	10	67 606	7	71 880	7	45 567	5	54 879	5	134 279	13	61 394	6	1 007 791	100

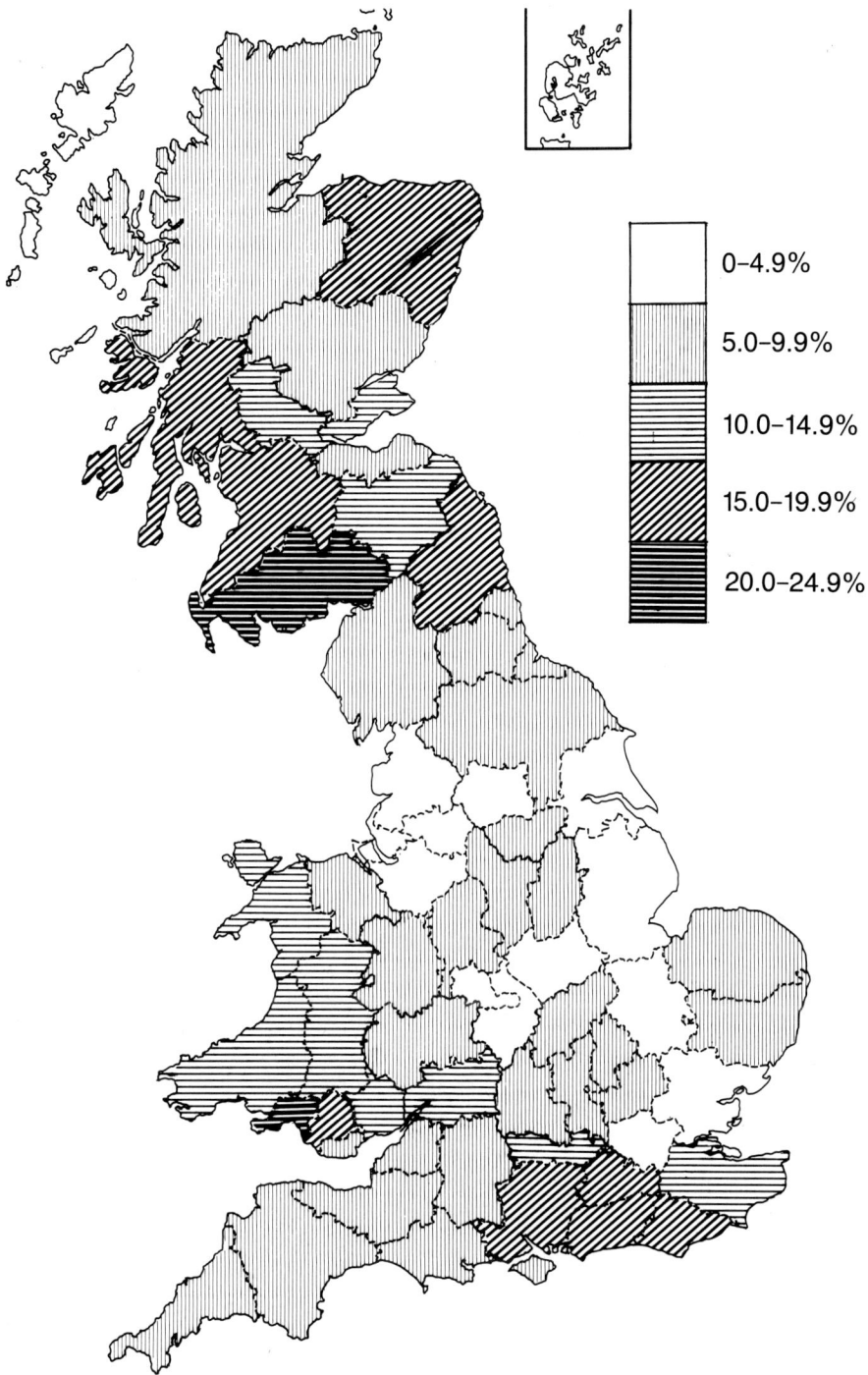

Figure 2. Woodland area as a percentage of land and inland water area. Counties in England and Wales; Regions in Scotland.

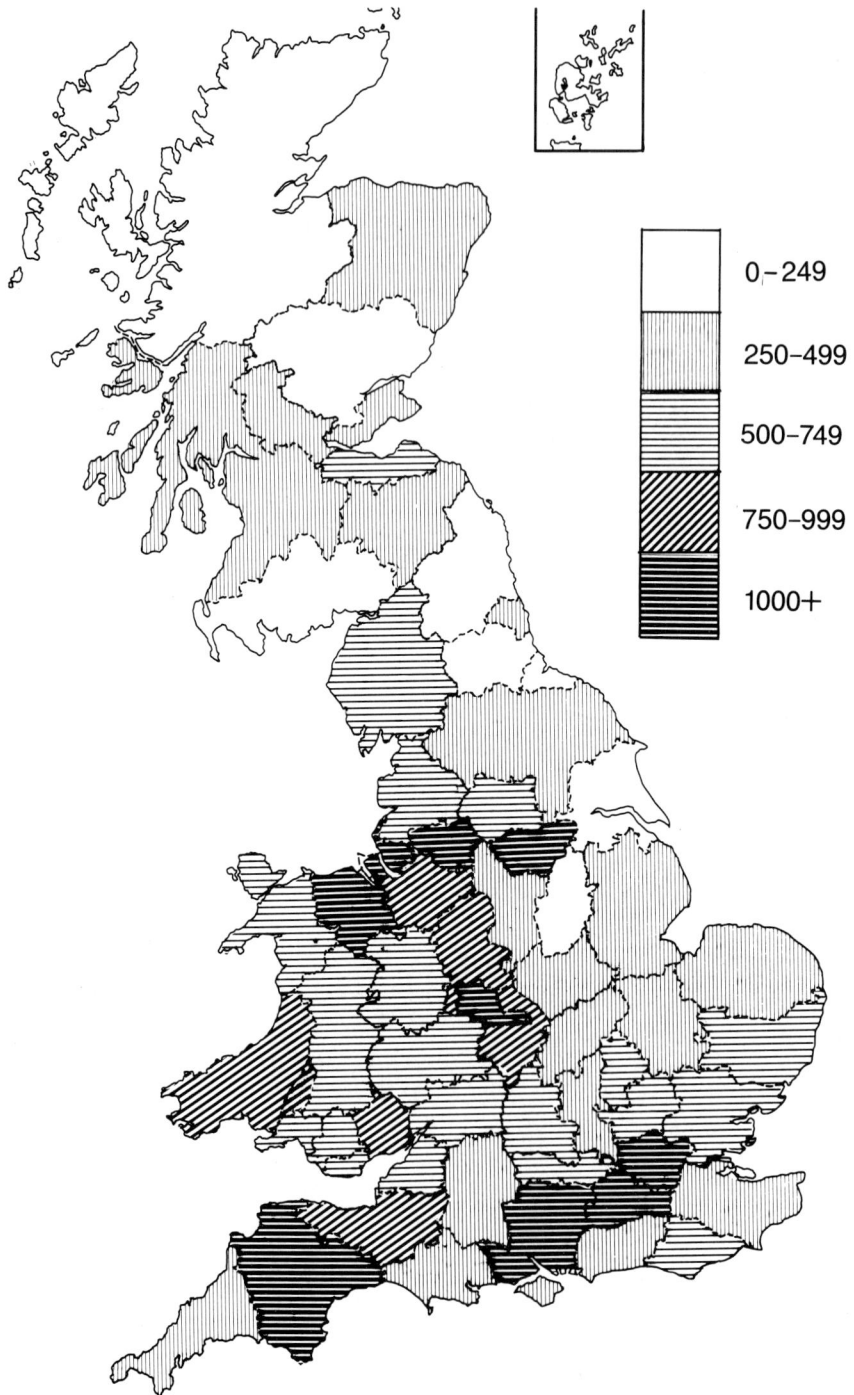

Figure 3. Number of live non-woodland trees per square kilometre. Counties in England and Wales; Regions in Scotland.

Figure 5. Broadleaved high forest. *(A10813)*

Figure 4. Coniferous high forest. *(A10814)*

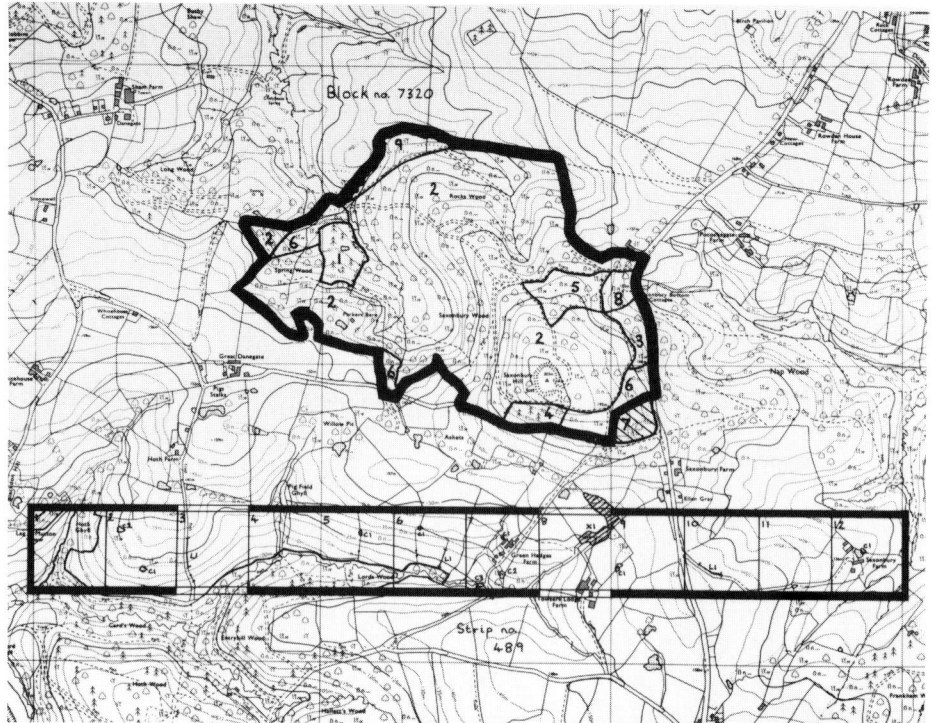

Figure 6. Map showing a woodland sample subdivided into its component blocks. Each block differs from its neighbour in terms of forest type, species, age, stocking, etc. Block 7 is an example of woodland extension.

An example of a non-woodland tree sample strip is also shown. All 12 squares are examined on aerial photographs and counts made of the number of isolated trees and the width and length of clumps and linear features. The two ground sample squares are numbers 3 and 8, with 8 containing an example of 'Extra' woodland. (*A10812* Ordnance survey base map, Crown copyright reserved)

Figure 7. Coppice with standards. Oak standards over Sweet Chestnut coppice. (*A10819*)

Figure 8.   Worked coppice. (*A10820*)

Figure 9.   Scrub hazel. (*A10816*)

Figure 10. An open-grown isolated tree. (*A10815*)

Figure 11. A clump. (*A10818*)

Figure 12.   A linear feature. (*A10821*)

Figure 13.   Ornamental trees in an urban area. (*A10822*)

Figure 14. An MSI 88 attached to an acoustic coupler for transfer of data over a telephone line. (*B9339*)

Figure 15. A CX400 being used for receipt and temporary storage of data. (*B9337*)

Table 14 Area of high forest by principal species and countries

Private woodlands

Hectares

Species	England Area	England % of category	England % of all species	Wales Area	Wales % of category	Wales % of all species	Scotland Area	Scotland % of category	Scotland % of all species	Great Britain Area	Great Britain % of category	Great Britain % of all species
Scots pine	50 885	28	9	1 221	3	1	83 072	29	24	135 178	27	13
Corsican pine	9 657	5	2	286	1	—	706	—	—	10 649	2	1
Lodgepole pine	3 210	2	1	1 040	3	1	21 427	8	6	25 677	5	3
Sitka spruce	18 652	10	3	19 485	51	22	108 535	38	31	146 672	29	14
Norway spruce	20 030	11	4	3 257	8	4	22 207	8	6	45 494	9	5
European larch	17 617	10	3	1 961	5	2	10 735	4	3	30 313	6	3
Jap./Hybrid larch	23 038	12	4	5 227	14	6	19 729	7	6	47 994	10	5
Douglas fir	12 144	7	2	3 159	8	4	5 190	2	1	20 493	4	2
Other conifers	7 631	4	1	1 077	3	1	1 840	1	1	10 548	2	1
Mixed conifers	21 002	11	4	1 336	4	1	7 352	3	2	29 690	6	3
Total conifers	183 866	100	33	38 049	100	42	280 793	100	80	502 708	100	50
Oak	114 442	31	20	23 758	45	26	15 309	21	4	153 509	31	15
Beech	41 242	11	7	3 208	6	4	9 933	14	3	54 383	11	5
Sycamore	34 369	9	6	3 704	7	4	9 062	13	3	47 135	9	5
Ash	53 677	14	9	9 018	17	10	3 953	5	1	66 648	13	7
Birch	42 726	11	8	5 409	10	6	15 846	22	4	63 981	13	6
Poplar	11 487	3	2	328	1	<1	306	<1	<1	12 121	2	1
Sweet chestnut	8 873	2	2	369	1	<1	7	<1	<1	9 249	2	1
Elm	5 334	1	1	355	1	<1	3 606	5	1	9 295	2	1
Other broadleaves	18 501	5	3	3 539	6	4	4 938	7	1	26 978	5	3
Mixed broadleaves	48 873	13	9	3 491	6	4	9 420	13	3	61 784	12	6
Total broadleaves	379 524	100	67	53 179	100	58	72 380	100	20	505 083	100	50
Total	563 390	100	100	91 228	100	100	353 173	100	100	1 007 791	100	100

Note: The total area of conifer and of broadleaved species in this and subsequent High Forest tables differs from the total area classified as Mainly conifer and Mainly broadleaved in Tables 12 and 13. Here areas have been allocated on the basis of species distribution and not forest type distribution.

E

Table 15 Area of high forest by principal species and planting year classes – Great Britain

Private woodlands

Hectares

Species	P71–80	P61–70	P51–60	P41–50	P31–40	P21–30	P11–20	P01–10	P1861–1900	Pre-1861	Totals
Scots pine	11 602	30 226	29 653	12 654	10 457	12 204	5 692	5 866	11 754	5 070	135 178
Corsican pine	1 849	4 315	1 563	394	937	575	389	250	313	64	10 649
Lodgepole pine	12 379	10 961	2 129	126	43	20	17	2	—	—	25 677
Sitka spruce	75 702	47 921	14 488	4 454	2 243	1 304	253	86	175	46	146 672
Norway spruce	7 978	18 878	9 234	3 626	1 863	2 149	749	407	515	95	45 494
European larch	2 271	4 730	4 962	3 670	4 193	3 502	1 885	2 549	2 220	331	30 313
Jap./Hybrid larch	8 965	13 887	14 777	4 234	2 880	2 355	547	211	121	17	47 994
Douglas fir	3 406	7 499	5 116	1 041	1 031	1 260	490	193	299	158	20 493
Other conifers	2 356	4 426	1 920	323	299	283	133	176	501	131	10 548
Mixed conifers	3 982	7 849	6 689	2 430	1 871	1 987	1 259	940	2 068	615	29 690
Total conifers	130 490	150 692	90 531	32 952	25 817	25 639	11 414	10 680	17 966	6 527	502 708
Oak	1 756	2 568	3 957	9 664	7 431	10 815	11 728	19 665	57 853	28 072	153 509
Beech	2 093	4 385	3 317	1 561	1 738	1 840	2 574	5 124	17 147	14 604	54 383
Sycamore	2 210	3 961	5 440	8 728	5 879	4 637	3 735	3 598	7 100	1 847	47 135
Ash	1 698	2 486	7 189	13 022	8 976	10 721	5 685	6 390	8 575	1 906	66 648
Birch	2 656	8 361	16 245	19 091	8 028	4 936	2 634	1 093	788	149	63 981
Poplar	1 887	3 985	4 054	974	276	513	299	37	74	22	12 121
Sweet chestnut	308	316	958	1 088	1 095	908	728	956	1 426	1 466	9 249
Elm	128	199	342	999	906	1 332	631	998	3 127	633	9 295
Other broadleaves	1 277	1 726	4 336	6 079	2 509	4 788	1 840	1 158	2 653	612	26 978
Mixed broadleaves	2 582	3 613	6 029	6 268	4 949	5 757	4 293	5 171	17 568	5 554	61 784
Total broadleaves	16 595	31 600	51 867	67 474	41 787	46 247	34 147	44 190	116 311	54 865	505 083
Total	147 085	182 292	142 398	100 426	67 604	71 886	45 561	54 870	134 277	61 392	1 007 791

larch are the two best represented species but there is also a substantial area classed as Mixed conifers, much of it in the form of amenity plantings in estate woodlands. Whilst these crops embrace a wide variety of species a substantial proportion of the total is still likely to be composed of the above two species.

All broadleaved species show a decline in the area planted or regenerated since the war but most of them show an increase around the P41–50 period with a rise in area of birch, ash, sycamore and, to a lesser extent, oak being particularly marked. This is likely to be due to regrowth from felled stands or colonisation of them. Most of the area in the older P Year classes is composed of oak and beech but sycamore and ash are quite prominent in the P1861–1900 class. There is also a substantial area classed as Mixed broadleaves covering a wide range of species of which the above four are again likely to be well represented.

The importance of individual species in each P Year class is shown in Table 16 for Great Britain.

The recent importance of Sitka spruce is highlighted here as is that of Lodgepole pine but, with the exception of birch in the P41–50 class, the rest of the P Year classes are dominated by Scots pine in the middle age crops and oak in the older. Broadleaved species, however, are second in importance in seven out of the ten P Year classes. The pattern differs somewhat between countries and is shown in Table 17.

Conifers dominate in only the two youngest P Year classes in England and in the three youngest in Wales, whereas in Scotland conifers form the principal species in all but the very oldest class. Oak dominates the older classes in both England and Wales with ash and/or birch forming the middle age classes. In Scotland, Scots pine is the predominant species in every class from 1861 to 1960.

Coppice and coppice with standards

These categories together account for 38 100 hectares or 3 per cent of the total private woodland area; 11 536 hectares or 30 per cent of the area of the category is composed of Coppice with Standards and 26 504 hectares or 70 per cent by Coppice. Almost all the Coppice with Standards and Coppice is in England. The distribution of the area of these two forest types is shown in Tables 18 and 19.

Table 16 Principal species in high forest by planting year classes – Great Britain

Private woodlands

Planting year classes	Principal species by percentage of area					
	First	%	Second	%	Third	%
P71–80	Sitka spruce	51	Lodgepole pine	8	Scots pine	8
P61–70	Sitka spruce	26	Scots pine	17	Norway spruce	10
P51–60	Scots pine	21	Birch	11	Jap./Hybrid larch	10
P41–50	Birch	19	Ash	13	Scots pine	13
P31–40	Scots pine	15	Ash	13	Birch	12
P21–30	Scots pine	17	Oak	15	Ash	15
P11–20	Oak	26	Scots pine	12	Ash	12
P01–10	Oak	36	Ash	12	Scots pine	11
P1861–1900	Oak	43	Mixed broadleaves	13	Beech	13
Pre 1861	Oak	46	Beech	24	Mixed broadleaves	9

Table 17 Principal species in high forest by planting year classes

Private woodlands

Planting year classes	England		Wales		Scotland		Great Britain	
	First	%	First	%	First	%	First	%
P71–80	Sitka spruce	17	Sitka spruce	66	Sitka spruce	65	Sitka spruce	51
P61–70	Scots pine	16	Sitka spruce	58	Sitka spruce	39	Sitka spruce	26
P51–60	Birch	16	Sitka spruce	24	Scots pine	37	Scots pine	21
P41–50	Birch	18	Oak	34	Scots pine	28	Birch	19
P31–40	Ash	16	Oak	22	Scots pine	33	Scots pine	15
P21–30	Oak	19	Ash	24	Scots pine	40	Scots pine	17
P11–20	Oak	26	Oak	51	Scots pine	43	Oak	26
P01–10	Oak	39	Oak	54	Scots pine	33	Oak	36
P1861–1900	Oak	48	Oak	60	Scots pine	30	Oak	43
Pre 1861	Oak	51	Oak	64	Oak	28	Oak	46

Table 18   Area of coppice by principal species – Great Britain

Private woodlands   Hectares

Sub type		Principal species of coppice						Total
		Sycamore	Ash	Sweet chestnut	Hornbeam	Hazel	Other species	
With standards	ha	119	193	5 254	1 697	1 464	2 809	11 536
	%	1	2	45	15	13	24	100
Coppice only	ha	2 379	1 554	12 866	1 716	1 606	6 443	26 564
	%	9	6	49	6	6	24	100
Total	ha	2 498	1 747	18 120	3 413	3 070	9 252	38 100
% of coppice total		7	5	47	9	8	24	100

Note: 'Other species' includes mixtures of the above five named species as well as other minor species of coppice.

Table 19   Area of coppice with standards by principal species of both coppice and standards – Great Britain

Private woodlands   Hectares

Principal species of standard	Principal species of coppice						Total
	Sycamore	Ash	Sweet chestnut	Hornbeam	Hazel	Other species	
Conifers	—	—	4	4	—	—	8
Oak	101	173	4 893	1 594	1 443	2 782	10 986
Ash	8	20	—	88	21	27	164
Sweet chestnut	—	—	353	—	—	—	353
Other broadleaves	10	—	4	11	—	—	25
Total	119	193	5 254	1 697	1 464	2 809	11 536
% of total	1	2	45	15	13	24	100

Table 18 shows that Sweet chestnut is the most important coppice species occupying 45 per cent of the area under Standards and 49 per cent of the Simple Coppice area. Hornbeam and hazel still have substantial areas under both Coppice types but sycamore and ash occur predominantly as Simple Coppice. Nearly a quarter of the total area in each case is composed of other species or mixtures of species which may include those already named. Intrusive species such as sycamore, ash and birch are often cut over at the same time as the main species and so help to form a mixed coppice crop. Mixtures of species are not commercially desirable where high value Sweet chestnut is concerned but mixed crops can be quite acceptable in other cases.

Where Standards are concerned almost all are oak. Standards of other species are a relatively rare occurrence and, apart from the coniferous species, may well have grown up by accident rather than by design.

Although Coppice and Coppice with Standards occur in most counties in England and Wales south of the Mersey-Humber line the main concentration is in Kent and to a lesser extent in East and West Sussex. These three counties hold about two-thirds of the Coppice with Standards and three-quarters of the Simple Coppice in the country. There is, however, an increasing interest now in coppice management throughout England and Wales and whilst it is unlikely that there will be a dramatic increase in the next few years in the area worked it is almost certain that the decline has been halted and some modest increase in the area of coppice types can be expected. The interest stems from two main reasons; first, the increasing demand for pole sized timber for wood burning stoves, agricultural and other such uses, and second the relatively short rotations and the usually variable age class structure of coppice woods produce habitats which are much more beneficial to flora and fauna than crops which have longer rotations and may include an extensive clear fell stage.

Scrub

The total area of Scrub is 142 482 hectares or 12 per cent of the total woodland area. Of this sum, England holds 77 488 hectares or 54 per cent, Wales 7205 hectares or 5 per cent and Scotland 57 789 hectares or 41 per cent. The distribution of species by countries is shown in Table 20.

From this Table it will be seen that in Great Britain birch is clearly the most important species occupying well over 40 per cent of the total. Of the named species oak is next in importance with 12 per cent followed by hazel, alder and ash with about 6 per cent each. The Other broadleaves category accounts for about one-fifth of the total and contains a wide variety of species but it is known that thorn,

mainly hawthorn, and sallow account for a large proportion of the area. The small area of conifers classed as Scrub are usually shelterbelts or scattered groups of trees at high elevation or in exposed conditions which are so windswept as to be incapable of producing much in the way of utilisable timber, if indeed, they are ever likely to be harvested at all.

When the position within individual countries is considered the pattern of species distribution is somewhat different from that in Great Britain. In England birch is rather less important, amounting to about one-quarter of the total, whereas oak and ash are more important. Wales, on the other hand, has oak and alder as the two major species with birch third followed closely by hazel. Birch is very clearly dominant in Scotland accounting for over two-thirds of the total with oak the only other species of any consequence.

It should be made clear that Scrub woodland is not necessarily an indicator of site potential. Some sites are inherently poor and can only grow poor crops, others may have reasonable soil conditions but exposure may be the reason why the crop has been downgraded. In other cases, however, the crop on the site may be composed of Scrub which is merely a stage of the succession from bare ground to woodland. Species such as hawthorn are often one of the first to colonise a site but once it has become established it tends to exclude stock. Species such as oak and ash can then often grow up in the shelter provided and what at one stage can look a fairly unpromising crop can, given time, provide the nucleus of a High Forest stand. Natural development

Table 20 Area of scrub by principal species

Private woodlands   Hectares

Principal species	England		Wales		Scotland		Great Britain	
	Area	%	Area	%	Area	%	Area	%
Conifers	707	1	—	—	1 514	3	2 221	2
Oak	12 379	16	1 249	17	3 928	7	17 556	12
Beech	766	1	36	1	248	—	1 050	1
Sycamore	1 903	2	82	1	364	—	2 349	2
Ash	7 197	9	218	3	372	1	7 787	5
Birch	19 970	26	1 126	16	41 052	71	62 148	44
Sweet chestnut	249	—	—	—	8	—	257	—
Alder	4 620	6	1 370	19	2 209	4	8 199	6
Hornbeam	398	1	10	—	—	—	408	—
Hazel	5 103	7	965	13	2 409	4	8 477	6
Willow	4 005	5	367	5	592	1	4 964	3
Other and mixed broadleaves	20 191	26	1 782	25	5 093	9	27 066	19
Total	77 488	100	7 205	100	57 789	100	142 482	100

too can have an effect in that many stands cut over during the war and then left to struggle on as best they could are now of a form and stocking worthy of being upgraded to a productive category although the marks of their origin are often still visible. There will always be a sizeable area classed as Scrub but not all of it will be permanently confined to this category.

Cleared

Cleared land is land which is at present bare or has only a low stocking of trees, but where the intention is to replant the area or to allow a regrowth from broadleaved coppice or felled stumps. At the present time the area in this category in each of the three countries forms about 2 per cent of their respective total woodland areas. The area in this class is likely to vary partly as a result of the age class structure of High Forest and partly to economic considerations. Because of the current imbalance in the age class structure the area of woodland classed as Cleared can be expected to rise relatively slowly until the major post-war plantings come up to rotation age towards the end of the century when a fairly dramatic increase can be expected. The category, however, is merely a stage in the life of successive tree crops and it is likely that most cleared areas will be restocked within two to four years after felling.

Mention needs to be made of two categories whose areas are not included in Table 12. They are Disforested and 'Extra' woodland. Although losses and gains from these causes can affect the area of woodland in all three ownerships the methods of assessment adopted in the Census meant that any necessary adjustments had already been made to the Forestry Commission and Dedicated and Approved woodland figures and so attention needed to be paid only to the 'Other' private woodland total.

Disforestation

Disforestation occurs when areas which have hitherto been classed as woodland are converted to some other form of land use. Such areas were identified in the Census when woodland blocks selected for sampling were inspected, either on an aerial photograph or on the ground, and parts or whole blocks were found to be no longer in existence. The area classed as Disforested in a county or Conservancy was then calculated using the procedure set out in Chapter 9. Any areas classed as Technically Disforested, i.e. areas which were shown as green on the 1:50 000 map but found on inspection not to meet the Census definition of woodland were dealt with in a similar fashion but are excluded from the following discussion.

The area of woodland in each county quoted as having been converted to other land uses is a function of the annual rate at which conversion has taken place and the period over which conversion has been measured. In the case of the Census this period relates to the date of the last major map revision. However, because of the very variable revision dates, and the fact that parts even of the same map can have differing dates, it is impossible to calculate the average annual rate of woodland loss from the existing data. All that the results can do is to indicate the land uses to which woodland has been converted.

In Great Britain the major cause of woodland loss is agriculture which accounts for about 64 per cent of the total. The percentage in England is similar to that of Great Britain, in Wales it is about 80 per cent and in Scotland about 55 per cent. Building is the next most important cause accounting for about 7 per cent of the overall loss, but with the England figure 10 per cent, followed by mining and quarrying with 5 per cent. Losses due to road construction and road widening account for only 0.5 per cent of the Great Britain total but is nearer 2 per cent in England. The balance of the total, which for Great Britain amounts to about 23 per cent, comprises a wide variety of causes including power lines, military installations, flooding for reservoirs, industrial sites, etc.

Extra woodland

This category has already been discussed in some detail in Chapter 7 but can briefly be defined as woodlands which were not marked as such on the 1:50 000 maps used for the Census but were found during the course of the Non-Woodland Tree Survey to be areas that would qualify as woodland under the definition then in force. For the reasons explained earlier it was decided not to include the area of this category in the Census because of the low precision that would inevitably be attached to the estimate and the lack of detailed knowledge about forest type, species, age class and so on. Nonetheless the area is substantial and is estimated to be about 88 000 hectares of which 52 000 were in England, 12 000 in Wales and 24 000 in Scotland. However, as with Disforested, it must be remembered that the area found during the course of the assessment is dependent on the date of the last major revision of the map base. This in some instances could be 20 years or more earlier and a considerable area of woodland could become established either naturally or artificially during the intervening period. It must also be said that much of this 'Extra' woodland is unmanaged and a considerable proportion of it could well disappear as a result of neglect, conversion to other land uses, fire and so on. Although it amounts at present to about 4 per cent of the total woodland area a proportion of it at least will survive to be included in the next Census survey. If the rate of change to woodland area is not to be distorted a more accurate method of determining the total area and characteristics of these woodlands will need to be found for the future.

Shrub layer

During the course of the Census information on species present in the shrub layer, if one was present, was collected on all stands that were ground assessed. As a consequence the data relate only to 'Other' private woodland.

As might be expected the presence of a shrub layer in stands of Coniferous High Forest was uncommon and occurred on less than 1 per cent of the area of this forest type in Great Britain. Elder was the dominant species but rhododendron and hazel were also present in significant quantities.

In Broadleaved High Forest hazel was the most widespread species occurring on 6 per cent of the area. The importance of this species probably results from many of the present High Forest stands having been originally of Coppice or Coppice with Standards origin. In the case of Scotland, however, rhododendron was the predominant species, and this species is also locally dominant in parts of England and Wales. Other species present in quantity are hawthorn and elder.

In both Coppice and Coppice with Standards 'Other shrubs' are classed as the dominant species group but specific mention is made of hawthorn, hazel, elder, willow and rhododendron as main supporting species. In Scrub and Cleared hazel, elder, sallow and hawthorn are the species most commonly present.

In no country does the shrub layer exceed 7 per cent of any individual forest type, although in individual counties, particularly in southern England, much higher percentages were recorded and in the case of Broadleaved High Forest exceeded 30 per cent.

CHAPTER 13

Forestry Commission woodland area

The distribution of the 891 715 hectares in Forestry Commission ownership is shown in Table 21. From this it will be apparent that Scotland holds 498 000 hectares or 56 per cent of the total, England has 255 000 hectares or 28 per cent and Wales 139 000 hectares or 16 per cent.

category amounting to only about 1 per cent of the total area. The Cleared areas are in practically all cases awaiting immediate restocking whilst the Scrub areas have generally been earmarked for retention and are unlikely to be specifically treated other than to thin them, where this is appropriate, and to ensure their continued existence.

Forest type

Ninety eight per cent of the area in all three countries and in Great Britain is composed of High Forest. Coniferous High Forest dominates all three countries and it is only in England that there is a significant proportion of broadleaved woodland amounting to some 17 per cent of the total. Virtually all the Coppice and Coppice with Standards is also in England. Scrub and Cleared woodland are relatively unimportant in all three countries, each

Planting year class

The age class pattern is shown in Table 22 from which it will be evident that the coniferous area in all three countries is heavily weighted towards the younger P Year groups. Sixty six per cent of the coniferous area in England has been planted since 1951 as has 74 per cent of the Welsh total and 83 per cent of the Scottish. All three countries have some representation in the P21–50 age classes but the area older than P21 is negligible in all cases.

Table 21   Area of woodland by forest type and country

Forestry Commission woodlands   Hectares

Forest type	England		Wales		Scotland		Great Britain	
	Area	%	Area	%	Area	%	Area	%
Mainly coniferous high forest	204 379	81	129 930	94	484 994	97	819 303	92
Mainly broadleaved high forest	43 976	17	6 123	4	4 033	1	54 132	6
Total high forest	248 355	98	136 053	98	489 027	98	873 435	98
Coppice with standards	32	<1	—	—	—	—	32	<1
Coppice	999	<1	1	<1	—	—	1 000	<1
Total coppice	1 031	<1	1	<1	—	—	1 032	<1
Scrub	2 010	1	1 017	1	2 723	1	5 750	1
Cleared	3 572	1	1 783	1	6 143	1	11 498	1
Total	254 968	100	138 854	100	497 893	100	891 715	100

Note: These figures do not agree exactly with those in the Forestry Commission's Annual Report for 1979–80 as a result of minor differences in classification.

Table 22 Area of high forest types by planting year classes

Forestry Commission woodlands

Hectares

		Planting year classes																				
Country	High forest type	P71–80		P61–70		P51–60		P41–50		P31–40		P21–30		P11–20		P01–10		P1861–1900		Pre 1861		Total	
		Area	%	Area	%	Area	%	Area	%	Area	%	Area	%	Area	%	Area	%	Area	%	Area	%	Area	%
England	Mainly coniferous	26 819	13	48 471	24	60 248	29	28 061	14	21 319	10	17 473	9	727	<1	498	<1	605	<1	158	<1	204 379	100
	Mainly broadleaved	1 225	3	3 307	8	11 463	26	7 015	16	5 923	13	3 155	7	1 126	3	788	2	3 236	7	6 738	15	43 976	100
	Total	28 044	11	51 778	21	71 711	28	35 076	14	27 242	11	20 628	8	1 853	1	1 286	1	3 841	1	6 896	2	248 355	100
Wales	Mainly coniferous	18 239	14	33 215	26	44 351	34	16 835	13	13 266	10	3 791	3	94	<1	39	<1	95	<1	5	<1	129 930	100
	Mainly broadleaved	125	2	385	6	2 380	39	650	11	532	9	305	5	383	6	73	1	1 004	16	286	5	6 123	100
	Total	18 364	13	33 600	25	46 731	34	17 485	13	13 798	10	4 096	3	477	<1	112	<1	1 099	1	291	<1	136 053	100
Scotland	Mainly coniferous	153 985	32	133 389	27	115 063	24	38 725	8	27 749	6	13 462	3	610	<1	374	<1	1 299	<1	338	<1	484 994	100
	Mainly broadleaved	381	9	192	5	740	18	466	12	610	15	250	6	145	4	49	1	881	22	319	8	4 033	100
	Total	154 366	32	133 581	27	115 803	24	39 191	8	28 359	6	13 712	3	755	<1	423	<1	2 180	<1	657	<1	489 027	100
Great Britain	Mainly coniferous	199 043	24	215 075	26	219 662	27	83 621	10	62 334	8	34 726	4	1 431	<1	911	<1	1 999	<1	501	<1	819 303	100
	Mainly broadleaved	1 731	3	3 884	7	14 583	27	8 131	15	7 065	13	3 710	7	1 654	3	910	2	5 121	2	7 343	14	54 132	100
	Total	200 774	23	218 959	25	234 245	27	91 752	11	69 399	8	38 436	4	3 085	<1	1 821	<1	7 120	<1	7 844	1	1 873 435	100

Table 23 Area of high forest by principal species and countries

Forestry Commission woodlands

Hectares

Species	England			Wales			Scotland			Great Britain		
	Area	Percentage of category	of all species	Area	Percentage of category	of all species	Area	Percentage of category	of all species	Area	Percentage of category	of all species
Scots pine	40 189	20	16	4 371	3	3	61 299	13	12	105 859	13	12
Corsican pine	30 555	15	12	3 407	3	3	2 640	1	1	36 602	4	4
Lodgepole pine	12 039	6	5	6 855	5	5	82 497	17	17	101 391	12	12
Sitka spruce	56 947	28	23	66 216	52	49	256 066	52	52	379 229	47	45
Norway spruce	23 469	12	10	15 384	12	11	32 500	7	7	71 353	9	8
European larch	4 245	2	2	634	<1	<1	5 222	1	7	10 101	1	1
Jap./Hybrid larch	12 704	6	5	18 234	14	13	32 417	7	7	63 355	8	7
Douglas fir	12 919	6	5	7 549	6	6	6 438	1	1	26 906	3	3
Other conifers	9 811	5	4	6 829	5	5	4 472	1	1	21 112	3	2
Mixed conifers	663	<1	<1	408	<1	<1	1 289	<1	<1	2 360	<1	<1
Total conifers	203 541	100	82	129 887	100	95	484 840	100	99	818 268	100	94
Oak	14 910	34	6	2 329	37	2	1 242	30	1	18 481	34	2
Beech	16 586	37	7	2 404	39	3	563	13	<1	19 553	35	3
Sycamore	1 835	4	1	129	2	<1	327	8	<1	2 291	4	<1
Ash	2 415	5	1	369	6	<1	149	4	<1	2 933	5	<1
Birch	3 175	7	1	174	3	<1	801	19	<1	4 150	8	1
Poplar	1 270	3	<1	176	3	<1	23	1	<1	1 469	3	<1
Sweet chestnut	578	1	<1	43	1	<1	1	<1	<1	622	1	<1
Elm	211	<1	<1	3	<1	<1	5	<1	<1	219	<1	<1
Other broadleaves	1 793	4	1	179	3	<1	177	4	<1	2 149	4	<1
Mixed broadleaves	2 041	5	1	360	6	<1	899	21	<1	3 300	6	<1
Total broadleaves	44 814	100	18	6 166	100	5	4 187	100	1	55 167	100	6
Total	248 355	100	100	136 053	100	100	489 027	100	100	873 435	100	100

Table 24 Area of high forest by principal species and planting year classes – Great Britain

Forestry Commission woodlands

Hectares

Species	P71–80	P61–70	P51–60	P41–50	P31–40	P21–30	P11–20	P01–10	P1861–1900	Pre-1861	Totals
Scots pine	3 776	18 312	37 956	14 984	12 408	15 401	552	551	1 552	367	105 859
Corsican pine	6 567	9 136	9 317	3 717	4 536	3 148	103	44	33	1	36 602
Lodgepole pine	36 418	39 279	23 429	896	1 042	326	1	—	—	—	101 391
Sitka spruce	133 153	105 967	75 982	38 129	20 315	5 531	114	24	9	5	379 229
Norway spruce	2 359	12 163	23 019	14 607	14 610	4 218	269	51	54	3	71 353
European larch	332	1 449	2 417	1 236	2 857	1 248	172	147	202	41	10 101
Jap./Hybrid larch	10 295	10 248	30 063	7 284	4 048	1 339	54	14	10	—	63 355
Douglas fir	3 091	8 306	9 100	1 786	1 503	2 896	96	51	36	41	26 906
Other conifers	2 530	9 271	7 356	807	692	368	35	9	29	15	21 112
Mixed conifers	378	646	897	107	144	94	14	32	41	7	2 360
Total conifers	198 899	214 777	219 536	83 553	62 155	34 569	1 410	923	1 966	480	818 268
Oak	472	295	3 360	2 028	1 992	1 318	769	478	3 112	4 657	18 481
Beech	417	1 928	7 740	3 016	2 230	660	151	108	951	2 352	19 553
Sycamore	173	414	480	276	440	245	102	52	87	22	2 291
Ash	24	220	426	502	878	430	141	85	172	55	2 933
Birch	120	459	949	1 250	731	431	124	10	64	12	4 150
Poplar	120	478	678	58	50	81	2	1	1	—	1 469
Sweet chestnut	46	50	185	59	68	70	41	17	54	32	622
Elm	28	20	14	46	42	28	20	14	5	2	219
Other broadleaves	244	177	402	471	297	238	74	34	124	88	2 149
Mixed broadleaves	249	155	498	501	515	358	240	88	568	128	3 300
Total broadleaves	1 893	4 196	14 732	8 207	7 243	3 859	1 664	887	5 138	7 348	55 167
Total	200 792	218 973	234 268	91 760	69 398	38 428	3 074	1 810	7 104	7 828	873 435

The distribution of broadleaved species in Great Britain is influenced largely by that of England. Some 22 per cent of the total is older than 80 years of age but the remaining area is mainly contained in the P Years 1931–60 with the peak in the P51–60 class. Since then the areas planted have declined quite markedly in England and Wales although Scotland shows a modest increase in the last decade.

Composition by principal species

Species distribution is shown in Table 23.

As might be expected Sitka spruce is by far the most important coniferous species in Great Britain, amounting to some 47 per cent, and it dominates the totals in all three countries. Scots pine is second but is closely followed by Lodgepole pine. Norway spruce and Japanese/Hybrid larch also have significant quantities. The pattern differs a little between countries. In England, Corsican pine takes the place of Lodgepole pine as third in importance whereas in Wales, where Scots pine is only of local importance, Japanese/Hybrid larch is second and Norway spruce third. In Scotland afforestation on the poorer soils in recent years has resulted in the extensive use of Lodgepole pine which is now second in importance to Sitka spruce with Scots pine third.

Of the broadleaved species beech and oak are of major importance together accounting for nearly 70 per cent of the Great Britain total. Of the remaining individual species only birch and ash are of significance. In England and Wales the order of representation is beech and oak followed by birch in England and ash in Wales; in Scotland the order is oak, birch, beech.

The analysis of the High Forest area in Great Britain by P Year class shown in Table 24 illustrates the age class pattern of the species in rather greater detail.

Sitka spruce has expanded in area in every decade since it was first used as a major afforestation species in the 1920s; Scots pine, on the other hand, was used substantially up to, and during, the last war but in the 1951–60 period showed a marked increase as the major planting programmes built up, often being used in mixture with Sitka spruce. Since that time, however, its use has declined, markedly so in the last 10 years, but its decline has been matched to some degree by the increase in the use of Lodgepole pine, both in mixture and pure, as being a much more promising species on the exposed and poor nutrient sites of the uplands, especially in Scotland. Corsican pine has had fluctuating fortunes but is now used fairly extensively in restocking programmes in forests on the sandy soils of areas such as Thetford. Japanese/Hybrid larch is now much less important than in the immediate post-war era and the use of European larch has declined dramatically

in the last decade. Norway spruce, again one of the more traditional coniferous species in the pre-war period has shown a major drop in area of recent years. It must be remembered, however, that new planting on sites that have not borne woodland in the recent past now tends to be on somewhat poorer sites than those afforested in the fifties and sixties and species such as Norway spruce would find it difficult to survive under such conditions far less produce an economic crop.

With broadleaved species beech has a very substantial proportion of its area in the P51–60 age class and marks a period when not only was it being used for underplanting coniferous crops but, because of the acquisition of a lot of felled and derelict sites from private owners in lowland Britain, it was also used either pure or in mixture with other broadleaved species. Whilst there is a substantial area of oak in the older age classes this species also shows the same increase in the P1951–60 class and again reflects its use in the replanting of some of the better sites in the lowlands. Few of the other species make a major single contribution to the total and the areas are generally distributed fairly evenly over the age range. Birch perhaps is the exception with a fairly substantial increase during the period of the last war, a point which has already been discussed elsewhere.

The importance of particular species in individual P Year classes is shown in Table 25.

Table 25 Principal species in high forest by planting year classes – Great Britain

Forestry Commission woodlands

| Planting year classes | Principal species by percentage of area | | | | | |
	First	%	Second	%	Third	%
P71–80	Sitka spruce	66	Lodgepole pine	18	Jap./Hybrid larch	5
P61–70	Sitka spruce	48	Lodgepole pine	18	Scots pine	8
P51–60	Sitka spruce	32	Scots pine	16	Jap./Hybrid larch	13
P41–50	Sitka spruce	42	Scots pine	16	Norway spruce	16
P31–40	Sitka spruce	29	Norway spruce	21	Scots pine	18
P21–30	Scots pine	40	Sitka spruce	14	Norway spruce	11
P11–20	Oak	25	Scots pine	18	Norway spruce	9
P01–10	Scots pine	30	Oak	26	European larch	8
P1861–1900	Oak	44	Scots pine	22	Beech	13
Pre 1861	Oak	59	Beech	30	Scots pine	5

Table 26 Principal species in high forest by planting year classes – Great Britain

Forestry Commission woodlands

Planting year classes	England		Wales		Scotland		Great Britain	
	First	%	First	%	First	%	First	%
P71–80	Sitka spruce	48	Sitka spruce	69	Sitka spruce	69	Sitka spruce	66
P61–70	Sitka spruce	21	Sitka spruce	56	Sitka spruce	57	Sitka spruce	48
P51–60	Sitka spruce	22	Sitka spruce	42	Sitka spruce	35	Sitka spruce	32
P41–50	Sitka spruce	32	Sitka spruce	48	Sitka spruce	47	Sitka spruce	42
P31–40	Scots pine	22	Sitka spruce	39	Sitka spruce	36	Sitka spruce	29
P21–30	Scots pine	48	Sitka spruce	28	Scots pine	37	Scots pine	40
P11–20	Oak	31	Oak	41	Scots pine	30	Oak	25
P01–10	Oak	32	Oak	47	Scots pine	46	Scots pine	30
P1861–1900	Oak	52	Oak	46	Scots pine	50	Oak	44
Pre 1861	Oak	61	Oak	89	Scots pine	45	Oak	59

Table 27 Area of coppice by principal species – Great Britain

Forestry Commission woodlands   Hectares

Sub type		Principal species of coppice						Total
		Sycamore	Ash	Sweet chestnut	Hornbeam	Hazel	Other species	
With standards	ha	—	—	21	—	1	10	32
	%	—	—	66	—	3	31	100
Coppice only	ha	1	—	950	—	24	25	1 000
	%	<1	—	95	—	2	3	100
Total	ha	1	—	971	—	25	35	1 032
% of coppice total		<1	—	94	—	3	3	100

Note: 'Other species' includes mixtures of the above five named species as well as other minor species of coppice.

Table 28 Area of coppice with standards by principal species of both coppice and standards – Great Britain

Forestry Commission woodlands   Hectares

Principal species of standard	Principal species of coppice						Total
	Sycamore	Ash	Sweet chestnut	Hornbeam	Hazel	Other species	
Conifers	—	—	12	—	—	—	12
Oak	—	—	4	—	1	10	15
Ash	—	—	—	—	—	—	—
Sweet chestnut	—	—	—	—	—	—	—
Other broadleaves	—	—	5	—	—	—	5
Total	—	—	21	—	1	10	32
% of total	—	—	66	—	3	31	100

Here the dominance of Sitka spruce in all the P Year classes since 1931 is clearly shown and, in the older classes, oak and Scots pine are the only two species represented. The use of pine either pure or in mixture with other species is shown by the fact that Lodgepole pine is second in importance in the two youngest age classes, whilst Scots pine was clearly used in the two decades following the war. The two spruces, Norway and Sitka, were major species in the inter-war years and again in the older classes oak and Scots pine are the two dominant species. Japanese/Hybrid larch only rises to third position in the P71–80 class and the P51–60 class. European larch and beech now only occur as limited areas in the older age classes.

The pattern of species importance is not very different between countries. Sitka spruce is naturally dominant in all the younger age classes and Scots pine and oak in the older, but it is only in Scotland that coniferous species predominate in all age classes.

Coppice and coppice with standards

The area of Coppice and Coppice with Standards in Forestry Commission ownership is small in relation to the total and is mainly confined to forests in south-east England (Tables 27 and 28).

The area of Coppice with Standards amounts to only 32 hectares with approximately half the area under oak standards and the bulk of the remainder under coniferous standards. Two-thirds of the coppice is composed of Sweet chestnut and the balance is composed of Other species, or is mixed. There are 1 000 hectares of Simple Coppice almost wholly composed of Sweet chestnut with only very minor quantities of hazel and other species. The coppice crops are currently worked on rotation and are likely to continue to be so as long as suitable markets for the produce exist.

Scrub

Scrub woodland (Table 29) amounts to 5 750 hectares or approximately 1 per cent of the total. A considerable area of it is recognised as being of scenic or conservation value and will be largely left untouched. Other areas may well be treated to encourage the existing better quality stems in order to bring the crops into a productive category whilst still retaining a largely unchanged outward appearance.

Scrub crops are not surveyed in detail and consequently a substantial part of the area is classed as Other broadleaves or as Mixed broadleaves. In many cases these crops are composed of undefined mixtures of species but it was not considered worthwhile to re-assess these crops to obtain more details of their crop constitution. It is likely, however, that much of this total is composed of birch, oak, alder and hazel, the same major species listed in those cases where the principal species were recorded.

Cleared

The area of cleared woodland, which in most cases is land from which the previous crop has been felled

Table 29 Area of scrub by principal species

Forestry Commission woodlands   Hectares

Principal species	England		Wales		Scotland		Great Britain	
	Area	%	Area	%	Area	%	Area	%
Conifers	117	6	5	1	101	4	223	4
Oak	189	9	133	13	176	6	498	9
Beech	9	—	2	—	9	—	20	—
Sycamore	8	—	—	—	9	—	17	—
Ash	79	4	6	1	4	—	89	2
Birch	420	21	2	—	690	25	1 112	19
Sweet chestnut	7	—	—	—	—	—	7	—
Alder	116	6	5	—	43	2	164	3
Hornbeam	2	—	—	—	—	—	2	—
Hazel	68	3	—	—	16	1	84	1
Willow	—	—	—	—	—	—	—	—
Other and mixed broadleaves	995	51	864	85	1 675	62	3 534	62
Total	2 010	100	1 017	100	2 723	100	5 750	100

and which is now awaiting re-stocking, amounts to 11 498 hectares, or about 1 per cent of the total. Just over half the area is in Scotland, just under a third in England and about 15 per cent in Wales. As the areas being clear felled annually increase so can an increase be expected in the area of this category. In normal circumstances, however, its size is unlikely to be more than two or three times the rate of felling current at the time.

CHAPTER 14

Non-woodland trees

General

The distribution of numbers of trees for individual countries by the categories of Isolated Trees, Clumps and Linear Features is shown in Table 30. From this it will be seen that England holds about 62.4 million trees of 7 cm diameter at breast height or over, or over 70 per cent of the total, and Wales and Scotland each have nearly 13 million trees, or about 15 per cent.

The largest number of trees is contained in Linear Features with 42 per cent of the total, followed by Clumps with 37 per cent and Isolated Trees with 21 per cent. However, the individual countries show somewhat differing distributions. England has approximately equal numbers of trees in Clumps and Linear Features, each accounting for nearly 40 per cent of the total; Scotland, again has fairly similar numbers in these two categories which together contain nearly 90 per cent of all trees. Wales, on the other hand, has over half the total number of trees in Linear Features, and nearly twice as many trees in Clumps than occur as Isolated Trees. The main point to bear in mind is that all three categories are essentially lowland in character; 85 per cent of the total tree numbers occur in England and Wales and these mainly south of the line drawn between the Mersey and the Humber.

The use of average tree density to compare one county with another can be misleading in that two counties can have similar averages but still have markedly different tree distribution patterns as a result of local variations in land use. Nevertheless the densities in some counties, especially the Metropolitan Counties, are surprisingly high as can be seen from the results in Appendix 8 and Figure 3 where Greater Manchester and West Midlands have tree densities as high as 1 400 trees per square kilometre. On the other hand, counties such as Durham and Northumberland and a Region like Highland, which have extensive upland areas, have relatively low values.

Alternatively one can use the area covered by Clumps and Linear Features as a basis for comparison and again, in some instances, the figure is surprisingly high. It must be remembered, however, that these areas represent measured crown area rather than ground area and consequently cannot be added to the woodland area totals directly. Nevertheless, even allowing for the fact that the areas are somewhat greater than they would be if measured in the conventional way the combined area comes to more than 200 000 hectares, most of which is broadleaved in character, and it is therefore not surprising that these lines of trees, avenues and small copses are so

Table 30   Number of live trees of 7 cm dbh or greater by category and country

Thousands of trees

Category	England		Wales		Scotland		Great Britain	
	Number	%	Number	%	Number	%	Number	%
Isolated trees	14 341	79	2 080	12	1 656	9	18 078	100
Clumps	23 461	72	3 855	12	5 328	16	32 644	100
Linear features	24 601	66	6 647	18	5 965	16	37 213	100
Total	62 403	71	12 582	14	12 949	15	87 935	100

Note: Figures may not total exactly due to rounding errors.

prominent in the lowland landscape and why, together with Isolated Trees their continued presence is so important to maintain the characteristics of the various regions.

Table 31 Area of clumps and linear features by type

		Hectares
Forest type	Clumps	Linear features
Mainly coniferous	7 411	5 218
Mainly broadleaved	91 333	111 486
Total	98 744	116 704

Table 32 shows for Great Britain the distribution of trees by category and by species and relates again to the number of live trees of 7cm dbh and over. Here the pattern of distribution is given by species and shows the relative importance of individual species in particular locations.

Conifers, for example, account for just over 10 per cent of the total but their representation in the Clump category is somewhat higher than this. Pines and cypresses occur most frequently in Clumps and the two genera together account for more than half the total number of coniferous trees. The importance of pines in the list is understandable but the high proportion of cypresses may occasion surprise. It is, however, now very much a genus of the cities and towns and is a common tree in gardens. If these trees are allowed to grow to their full stature their presence is likely to have a significant effect on the

Table 32 Number of live trees of 7cm dbh or greater by category and principal species

Thousands of trees

Species	Isolated trees		Clumps	Linear features	Total
	Boundary	Open			
Pines	93	355	1 321	687	2 456
Spruces	45	151	512	1 069	1 777
Larches	26	138	574	544	1 282
Cypresses	32	680	1 048	630	2 390
Other conifers	32	374	587	411	1 404
Total conifers	228	1 698	4 042	3 341	9 309
Oak	2 498	1 066	3 399	5 302	12 265
Beech	294	265	1 425	2 547	4 531
Sycamore	653	979	3 569	4 110	9 311
Ash	2 702	983	6 012	5 895	15 592
Birch	205	913	4 550	2 386	8 054
Poplar	133	333	566	946	1 978
Sweet chestnut	19	16	110	337	482
Horse chestnut	56	232	226	244	758
Alder	238	200	2 814	6 568	9 820
Lime	60	376	377	519	1 332
Elm	183	189	1 062	1 696	3 130
Willow	272	418	1 407	1 480	3 577
Other broadleaves	474	2 395	3 085	1 842	7 796
Total broadleaves	7 787	8 365	28 602	33 872	78 626
Total	8 015	10 063	32 644	37 213	87 935

Note: In addition, within Great Britain there are the following trees:

Species	Isolated trees	Clumps	Linear features	Total
Trees <7cm dbh. All species	9 857	5 572	5 090	20 519
Dead and dying. All species	1 494	1 591	2 139	5 224

F

future treescape of our urban areas particularly in the south of England where the use of cypress has been widespread. The genus undoubtedly has advantages in the short term for use in newly established housing estates and other situations where quick vegetative growth is a desirable attribute. The spruces, on the other hand, occur more frequently in Linear Features where this genus is regularly used for the creation of shelterbelts where their importance lies in their ability to withstand exposure.

When broadleaved trees are considered it will be seen that ash is the most numerous species in the country with over fifteen million trees and is followed by oak with twelve million, alder with ten million, sycamore with nine million and birch with eight million. These five species thus account for 70 per cent of the broadleaved total. Ash is most prevalent in Clumps and Linear Features with fairly equal numbers in each whilst oak occurs more frequently in Linear Features than Clumps. Alder is essentially a riparian tree and as a consequence occurs naturally in the form of long, narrow belts. Sycamore has the same distribution pattern as ash but birch occurs most frequently in the form of Clumps.

Isolated Trees are probably less frequent in their occurrence than might have been imagined but this is partly due to the method of classification whereby, because of the use of aerial photography in determining tree numbers and areas, two trees with their crowns touching had to be defined as a Clump rather than as two individual stems. Isolated Trees are distributed fairly equally between boundary trees, i.e. those within 2 metres of a boundary and open grown trees. Boundaries were, however, not recognised in urban areas so that all trees in urban areas were classed as being open grown. This may distort the relative values somewhat but nonetheless certain species dominate the Isolated Tree category. Oak has nearly 30 per cent of its total numbers classed as Isolated with nearly two and one-half times as many boundary trees as classed as open grown. It is therefore more commonly a tree of the hedgerows. Ash is also important with nearly one-quarter of its total classed as Isolated Trees and about three times as many boundary trees as are open grown. Few of the other individual species with the possible exception of sycamore and birch are numerically important in this category although species like Horse chestnut

Table 33 Number of all live trees by principal species and size classes

Thousands of trees

Species	Size class (dbh)					Total
	<7 cm	7–20 cm	21–30 cm	31–50 cm	>50 cm	
Pines	1 773	742	615	875	224	4 229
Spruces	3 476	1 364	234	141	38	5 253
Larches	206	420	370	440	52	1 488
Cypresses	2 310	2 089	196	89	16	4 700
Other conifers	508	641	294	289	180	1 912
Total conifers	8 273	5 256	1 709	1 834	510	17 582
Oak	747	2 979	1 820	3 412	4 054	13 012
Beech	214	1 457	867	1 124	1 083	4 745
Sycamore	1 687	4 943	1 564	1 778	1 026	10 998
Ash	1 567	7 987	2 385	3 247	1 973	17 159
Birch	1 610	6 109	1 254	620	71	9 664
Poplar	503	921	397	438	222	2 481
Sweet chestnut	15	253	108	59	62	497
Horse chestnut	163	148	117	194	299	921
Alder	715	4 943	2 605	1 986	286	10 535
Lime	157	366	254	407	305	1 489
Elm	180	1 522	559	640	409	3 310
Willow	765	1 845	537	482	713	4 342
Other broadleaves	3 923	5 484	1 231	862	219	11 719
Total broadleaves	12 246	38 957	13 698	15 249	10 722	90 872
Total	20 519	44 213	15 407	17 083	11 232	108 454

and lime do have a substantial proportion of their numbers as Isolated Trees. Elm at one time would have occasioned more than just a passing reference in this category but it now occupies less than 4 per cent of the broadleaved total and sadly its importance is likely to decline further. It occurs more frequently in Clumps and Linear Features than as an Isolated Tree.

The inclusion of trees under 7 cm dbh makes a substantial difference to the total increasing it by 20.5 million trees, from 87.9 to over 108 million. Nearly half these small trees were classed as Isolated Trees with the remainder divided reasonably equally between Clumps and Linear Features. Table 33 shows the effect that the inclusion of these trees under measurable size has on the species distribution and also shows the full diameter class distribution. It must be remembered, however, that although only trees which were considered to have a good chance of growing on have been included in the under 7 cm class it is a vulnerable category and its losses are likely to be heavier than in the larger size classes.

When species are considered it will be seen that the inclusion of this smallest class results in conifers nearly doubling their numbers with spruces showing the greatest gain. Shelterbelts are usually fenced and consequently relatively small trees are eligible in these cases and they probably account for the major part of the sizeable increase. Cypresses nearly double in numbers with garden trees likely to form the greater part of the total and while pines also show a major rise, larches show little change in numbers. Of the broadleaved species oak shows only a small increase and the species with the major gains are sycamore, birch and ash and also the Other broadleaves category which is likely to contain a wide variety of species, many of them ornamentals which are being used for amenity planting.

When size class distribution is considered it will be seen from Table 33 that conifers are heavily weighted towards the smaller size classes. Nearly half the total is under 7 cm dbh and 77 per cent of the total is 20 cm dbh or under. Only the pines, larches and Other conifers have numbers in quantity in the larger size classes.

Tree numbers in broadleaves show a more even distribution with 13 per cent under 7 cm dbh, 43 per cent in the 7–20 cm class, 15 per cent in the 21–30 cm class and 29 per cent in the two largest size classes. The distribution by species, however, shows some marked differences. Ash, for example, has a reasonable spread of trees throughout most of the size classes but is heavily weighted to the smaller end by having nearly half its total in the 7–20 cm dbh class. Oak, on the other hand, has more trees in the largest size class than in any other and the under

7 cm dbh class has the least. Unless measures are taken oak will decline in importance in future. Sycamore has a pattern similar to ash with heavy weighting to the 7–20 cm class as have birch and alder, but while sycamore does have quite a sizeable number of trees in the largest class both birch and alder, by virtue of their shorter life span, have relatively few trees larger than 30 cm in the case of birch and 50 cm in the case of alder. Beech has a more even distribution of tree numbers than most but is again short of numbers in the smallest class.

Appendix 11 gives information on the rate at which broadleaved trees can be expected to grow through the various size classes and therefore some indication of the changes that can be expected in the future pattern of diameter distribution in this category.

Health

Health was recorded for all live measurable trees of 7 cm dbh and over and three classes were recognised which are described in greater detail in Appendix 10. Good health was defined as showing no obvious signs of crown, bole or root damage, moderate health where there was a symptom of damage in one of the above areas and poor health where two or more symptoms were evident. A certain amount (up to 10 per cent) of stagheadedness was permitted in oak as this is not of itself necessarily a sign of ill-health, but dead branches or dead bark in beech automatically resulted in the tree being classed as in poor health. Elm was treated differently from other species and special checks were made for evidence of elm disease. Trees that were in an advanced stage of the disease were classed as dead in the expectation that they would have succumbed by the following Spring.

In general the health of the non-woodland tree population is good with the smaller size classes, as expected, being healthier than the older classes. Where broadleaved species are concerned England has 72 per cent classed as good, 17 per cent as moderate, 3 per cent as poor and 8 per cent as dead or dying, percentages which are almost identical with the Great Britain average. Wales, on the other hand, has 85 per cent of its tree numbers in good health as opposed to only 61 per cent in Scotland where the higher levels of exposure have an undoubted effect. Nonetheless, well over 90 per cent of all trees in all three countries are classed as having good or moderate health. Larch is the coniferous species with the poorest health record, many of the trees being in the small and middle range size classes. Elm obviously formed most of the dead and dying category and also a substantial part of the

Table 34 Health of all trees of 7 cm dbh or greater by species group and size class

Thousands of trees

Health class	England				Wales				Scotland				Great Britain			
	Conifer	%	Broadleaved	%	Conifer	%	Broadleaved	%	Conifer	%	Broadleaved	%	Conifer	%	Broadleaved	%
Good	4 541	85	44 367	72	1 280	86	9 632	85	2 235	84	6 401	61	8 050	84	60 402	72
Moderate	574	11	10 679	17	134	9	1 341	12	297	11	3 492	34	1 006	11	15 510	19
Poor	133	2	2 109	3	33	2	162	1	85	3	439	4	253	3	2 714	3
Dead or dying	118	2	4 645	8	43	3	244	2	52	2	123	1	214	2	5 010	6
Total	5 366	100	61 800	100	1 490	100	11 379	100	2 669	100	10 455	100	9 523	100	83 636	100

Note: Figures may not total exactly due to rounding errors.

poor category but ash, particularly in the Midlands, with most diameter classes involved, forms over a quarter of the tree numbers in the poor category. Also prominent are oak and willow in the over 50 cm class where age is obviously a major factor.

Some measure of how long the current landscape can be expected to last, given no interference from man, can be gauged by using the present health status for a county and comparing it with the Table of Life Expectancy given in Appendix 10. Obviously the latter is a national table and cannot be expected to give accurate predictions on a local basis but if, for example, it is considered that the national position overstates the local life expectancy by 10 per cent then the Table can be redrawn in the light of the local knowledge. Naturally other factors come into play but it can help in assessing when major changes in the landscape are likely to take place as a consequence of old age and ill-health taking its toll of the tree population.

CHAPTER 15

Standing volume

The methods of volume calculation described in Chapter 9 were used to produce the following estimates of standing volume. However, it must be made clear that no allowances have been made in these estimates for any constraints arising from planning, conservation or amenity considerations and consequently it must not be assumed that all the volume shown in the Tables is available for harvesting.

Woodlands

The volume standing in woodlands in Great Britain at 31 March 1980 was estimated to be 197.4 million cubic metres overbark. Of this total 106.3 million cubic metres was in Coniferous High Forest, 90.3 million in Broadleaved High Forest and 0.8 million in Standards over Coppice. No volume estimates were made for Scrub or Coppice. Table 35 shows the distribution of volume between countries and ownership classes.

From this Table it will be seen that while Forestry Commission woodlands account for over half the coniferous volume, the major part of the broadleaved volume is in private hands. Scotland holds slightly more coniferous volume than England but in both countries the volume is distributed fairly evenly between Forestry Commission and privately owned woodlands. This, however, is probably a temporary situation as the age class structure of the two ownerships is such that Forestry Commission woodland volume is likely to increase at a much faster rate than that of private woodlands for quite some time to come. Broadleaved volume, not surprisingly, is largely contained in England which holds well over 50 per cent of the volume in the combined High Forest categories, and virtually all the volume in Standards over Coppice.

The distribution of volume by species and P Year classes is shown for Great Britain in Table 36.

This Table shows that the volume of Scots pine, the most important species in volume terms in 1965, is now exceeded by that of Sitka spruce and the latter can be expected to increase markedly as the extensive area afforested with this species begins to come into the volume bearing category. At this stage of development, although the standing volume per hectare is low, the percentage volume increments are high and, when applied to the very substantial areas involved, this further results in large quantities being recruited annually to the volume bearing category. Currently most of the Sitka spruce volume is concentrated in crops in age classes P21–30 and younger whereas the Scots pine volume is spread more uniformly over the age range but with the greatest volume in the P21–30 class, the period that marked the start of the major upland afforestation programmes of the Forestry Commission. Two other species with substantial volumes are Norway spruce and Japanese/Hybrid larch.

Three species dominate the broadleaved volume, oak with 36 per cent, beech with 17 per cent and ash with 11 per cent. Birch, despite its considerable importance in area terms, does not figure prominently in volume terms, because of its relatively young average age and relatively low production capability. Much of the oak and beech volume is located in the two oldest P Year classes whereas that of ash is more evenly distributed over the age range. Sycamore also has quite a high proportion of its volume in the P1861–1900 class but most is concentrated in the P Year classes covering the inter-war period and the decade which followed it.

So far as the individual countries are concerned Scotland holds over half the Sitka spruce volume and England and Wales a little under a quarter each, whereas with Scots pine, whilst Scotland again holds about half the total, most of the remainder is in England. With oak, on the other hand, over three-quarters of its volume is in England which also holds nearly three-quarters of the volume of beech. Scotland holds nearly a quarter of the birch volume and Wales over 15 per cent of the ash.

The volume distribution can also be considered in terms of diameter class distribution and this is shown in Table 37.

About half the coniferous volume is in the 7–20 cm dbh class and the proportions in the other three classes are 27 per cent in the 21–30 cm class, 19 per cent in the 31–50 cm class and 4 per cent in the over

Table 35   Standing volume of timber by category and ownership

Millions of cubic metres overbark

Volume category	Private woodlands				Forestry Commission woodlands				Total woodlands			
	England	Wales	Scotland	Great Britain	England	Wales	Scotland	Great Britain	England	Wales	Scotland	Great Britain
In high forest												
Coniferous timber	21.1	3.1	22.1	46.3	21.7	12.7	25.6	60.0	42.8	15.8	47.7	106.3
Broadleaved timber	63.9	9.8	12.3	86.0	3.4	0.5	0.4	4.3	67.3	10.3	12.7	90.3
Total high forest	85.0	12.9	34.4	132.3	25.1	13.2	26.0	64.3	110.1	26.1	60.4	196.6
Timber in standards over coppice	0.8	—	—	0.8	—	—	—	—	0.8	—	—	0.8
Total	85.8	12.9	34.4	133.1	25.1	13.2	26.0	64.3	110.9	26.1	60.4	197.4

81

Table 36 Standing volume of timber in high forest and standards over coppice by principal series and planting year classes – Great Britain

All woodland ownerships

Thousands of cubic metres overbark

Species	P61–70	P51–60	P41–50	P31–40	P21–30	P11–20	P01–10	P1861–1900	Pre-1861	Total
Scots pine	1 140.9	4 414.5	3 378.3	4 333.6	6 261.9	1 570.6	1 710.7	3 386.5	1 500.6	27 697.6
Corsican pine	718.4	1 413.6	824.6	1 416.4	1 186.9	183.6	105.3	150.5	32.4	6 031.7
Lodgepole pine	763.6	1 521.0	119.1	163.9	53.2	4.9	0.7	–	–	2 626.4
Sitka spruce	3 928.6	9 385.2	7 132.0	5 378.8	2 080.6	141.7	45.1	75.9	19.1	28 187.0
Norway spruce	786.1	2 967.6	2 995.4	3 622.9	1 692.2	328.4	150.2	302.3	46.5	12 891.6
European larch	237.4	632.4	784.6	1 357.3	1 009.2	501.6	627.7	626.1	97.9	5 874.2
Jap./Hybrid larch	1 633.7	5 295.2	2 013.9	1 470.3	915.6	158.3	68.5	57.8	5.0	11 618.3
Douglas fir	879.5	1 749.5	593.4	773.7	1 494.3	224.9	107.1	172.5	91.0	6 085.9
Other conifers	397.9	945.7	224.8	278.9	122.0	42.9	57.3	153.8	51.7	2 275.0
Mixed conifers	49.6	331.0	250.3	346.4	474.5	314.6	254.7	791.6	179.5	2 992.2
Total conifers	10 535.7	28 655.7	18 316.4	19 142.2	15 290.4	3 471.5	3 127.3	5 717.0	2 023.7	106 279.9
Oak	29.4	341.6	1 383.5	1 408.8	1 996.0	2 226.4	3 814.4	14 210.6	7 483.1	32 893.8
Beech	34.3	396.6	362.3	512.6	465.0	590.9	1 438.4	5 868.1	5 551.6	15 219.8
Sycamore	233.0	551.7	1 350.5	1 119.8	931.2	812.9	759.0	1 718.8	488.4	7 965.3
Ash	135.6	696.4	1 627.0	1 358.5	1 966.8	1 105.6	1 135.8	1 680.2	534.0	10 239.9
Birch	386.3	1 365.6	2 170.2	894.2	636.0	306.7	192.5	134.3	23.3	6 109.1
Poplar	243.7	538.6	168.9	70.7	117.4	75.7	8.4	22.1	6.0	1 251.5
Sweet chestnut	7.3	97.4	224.0	194.4	217.4	167.3	284.0	479.1	685.1	2 356.0
Elm	5.0	34.2	151.5	208.9	376.8	131.5	251.3	880.8	175.5	2 215.5
Other broadleaves	84.0	508.1	734.2	451.3	823.5	333.9	244.5	736.6	147.5	4 063.6
Mixed broadleaves	61.4	335.2	621.6	625.6	885.0	714.2	992.4	3 488.1	1 118.2	8 841.7
Total broadleaves	1 220.0	4 865.4	8 793.7	6 844.8	8 415.1	6 465.1	9 120.7	29 218.7	16 212.7	91 156.2
Total	11 755.7	33 521.1	27 110.1	25 987.0	23 705.5	9 936.6	12 248.0	34 935.7	18 236.4	197 436.1

Table 37 Standing volume of timber in high forest and standards over coppice by principal species and size classes – Great Britain

All woodland ownerships   Thousands of cubic metres overbark

| Species | Size class (dbh) | | | | Total |
	7–20 cm	21–30 cm	31–50 cm	>50 cm	
Scots pine	10 063.7	8 148.0	7 602.1	1 883.8	27 697.6
Corsican pine	2 592.0	1 883.4	1 363.3	193.0	6 031.7
Lodgepole pine	2 519.7	99.4	7.3	—	2 626.4
Sitka spruce	18 643.1	6 764.4	2 598.2	181.3	28 187.0
Norway spruce	7 469.1	3 679.9	1 576.2	166.4	12 891.6
European larch	1 259.0	2 105.3	2 268.3	241.6	5 874.2
Jap./Hybrid larch	5 386.0	4 100.3	2 032.0	100.0	11 618.3
Douglas fir	2 525.0	984.4	1 858.8	717.7	6 085.9
Other conifers	1 409.2	403.2	311.1	151.5	2 275.0
Mixed conifers	1 115.5	775.0	682.9	418.8	2 992.2
Total conifers	52 982.3	28 943.3	20 300.2	4 054.1	106 279.9
Oak	2 803.2	5 024.6	13 875.3	11 190.7	32.893.8
Beech	1 374.0	1 113.7	4 554.8	8 177.3	15 219.8
Sycamore	1 372.5	1 860.0	3 311.3	1 421.5	7 965.3
Ash	2 142.2	2 755.8	3 877.5	1 464.4	10 239.9
Birch	3 115.9	1 868.1	1 020.2	104.9	6 109.1
Poplar	89.8	303.8	738.9	119.0	1 251.5
Sweet chestnut	277.7	306.0	622.5	1 149.8	2 356.0
Elm	219.3	308.2	974.4	713.6	2 215.5
Other broadleaves	1 048.3	1 073.7	1 130.9	810.7	4 063.6
Mixed broadleaves	1 347.3	2 725.0	2 241.4	2 528.0	8 841.7
Total broadleaves	13 790.2	17 338.9	32 347.2	27 679.9	91 156.2
Total	66 772.5	46 282.2	52 647.4	31 734.0	197 436.1

50 cm class. The relative immaturity of the Sitka spruce crops can be gauged by the fact that about two-thirds of the volume of this species is in the smallest class whilst in the case of Scots pine it is only a little over a third. Nonetheless it is a pattern which is changing very rapidly and whilst felling practice is likely to restrict the amount of volume moving into the over 50 cm dbh class the two middle size classes are likely to show substantial increases over the next few years, as indeed is the smallest class as the result of recruitment.

Broadleaved crops, on the other hand, show a much heavier weighting to the larger size classes. Here only 15 per cent of the volume is in the 7–20 cm class, 19 per cent in the 21–30 cm class, 35 per cent in the 31–50 cm class and 31 per cent in the over 50 cm class. Oak and beech are heavily weighted towards the larger size classes whereas ash and sycamore have their volumes fairly evenly spread over the range; birch, however, tends to have its volume concentrated in the smaller size classes.

The next few years are likely to show radical changes in the volume distributions, not only in terms of species, with Sitka spruce becoming more important and Japanese/Hybrid larch and Norway spruce becoming less so, but also in the relative proportions held by private owners and the Forestry Commission.

Non-woodland trees

The estimated standing volume in non-woodland trees is shown in Table 38 to be almost 28 million cubic metres overbark of which nearly 3 million are coniferous and 25 million broadleaved. England accounts for almost 75 per cent of the total volume, Wales for 14 per cent and Scotland for 11 per cent.

The distribution of volume between size classes shows there to be 4 per cent in the smallest size class, 7 per cent in the 21–30 cm class, 24 per cent in the 31–50 cm class and 65 per cent in the largest, the over 50 cm dbh class. This distribution pattern is one that is changing all the time as volume is lost from the larger size classes by death, felling or windblow

Table 38 Standing volume of timber for live trees of 7 cm dbh or greater by principal species and size classes – Great Britain

Thousands of cubic metres overbark

Species	Size class (dbh)				Total
	7–20 cm	21–30 cm	31–50 cm	>50 cm	
Pines	33.1	139.2	602.8	446.9	1 222.0
Spruces	60.2	56.9	75.8	56.1	249.0
Larches	31.2	85.9	233.9	96.5	447.5
Cypresses	74.7	37.7	45.0	31.2	188.6
Other conifers	32.8	71.7	157.8	616.9	879.2
Total conifers	232.0	391.4	1 115.3	1 247.6	2 986.3
Oak	72.1	195.4	1 375.9	7 175.5	8 818.9
Beech	38.9	118.9	494.1	2 046.1	2 698.0
Sycamore	115.2	221.2	696.3	1 548.3	2 581.0
Ash	195.3	359.4	1 316.2	2 265.7	4 136.6
Birch	82.3	103.5	126.1	56.6	368.5
Poplar	41.7	91.8	275.7	482.5	891.7
Sweet chestnut	20.8	26.5	30.5	292.2	370.0
Horse chestnut	2.9	15.1	91.3	574.3	683.6
Alder	129.6	370.5	610.3	310.3	1 420.7
Lime	8.3	33.6	201.5	724.9	968.3
Elm	24.9	63.7	229.1	641.7	959.4
Willow	44.7	59.3	117.4	347.3	568.7
Other broadleaves	25.1	43.6	135.1	294.6	498.4
Total broadleaves	801.8	1 702.5	5 699.5	16 760.0	24 963.8
Total	1 033.8	2 093.9	6 814.8	18 007.6	27 950.1

and is replaced by volume from trees in the smaller classes growing through to the larger, but the relativities are unlikely to change radically in the short term.

Generally the larger the diameter of the tree the greater its volume is likely to be and this naturally leads to a heavy weighting of volume towards the larger size classes. Where there is, in addition, a relatively high proportion of tree numbers in these larger size classes the volume can rise very substantially. For example, of the coniferous species only pines and the larches have a substantial proportion of their tree numbers in the 21–30 and 31–50 cm dbh classes and these two species groups account for over half the coniferous volume.

Oak has 35 per cent of the overall broadleaved volume, ash has 17 per cent and beech and sycamore about 10 per cent. Elm, second to oak in volume terms in 1951, is now only seventh with 4 per cent. In the case of both oak and beech many of the trees are of large size so that over 80 per cent and over 75 per cent respectively of the volume of these two species occurs in the over 50 cm class. In both instances there is likely to be a lack of volume being recruited to the larger classes so that the total

volume cannot be maintained in perpetuity.

Not all trees, of course, have utilisable volume, the proportion varying from county to county and region to region. Tree numbers present in a given area are therefore not necessarily an accurate measure of the amount of standing timber and estimates of the percentage of trees that have utilisable stem lengths vary from about 5 per cent in North Scotland Conservancy to 85 per cent in Counties such as Bedfordshire and Gloucestershire.

It is interesting to look at the volume in non-woodland trees as a proportion of the total timber volume in the country, i.e. from woodland and non-woodland sources alike. In the case of conifers the 3 million cubic metres in non-woodland trees represents less than 3 per cent of the total coniferous volume resource whereas with broadleaved species the 25 million cubic metres in non-woodland trees account for over 20 per cent of the broadleaved total. The importance of the broadleaved non-woodland trees in terms of the total standing volume is therefore clearly evident and even more so when it is appreciated that nearly 40 per cent of the volume in trees of over 50 cm dbh occur outside of woodland.

CHAPTER 16

Volume increment

The estimates of standing volume presented in Chapter 15 then formed the basis upon which the calculations of increment were made.

Current annual increment is the amount of timber volume which is added to the timber stocks each year as a result of natural growth and that figure, after allowance has been made for natural losses such as fire, disease and windsnap, can be used to determine sustainable levels of felling.

The rate of annual increment differs according to species, age and stocking of individual crops but, in general, increment in conifers tends to be laid down at its fastest rate between the ages of 25 and 40 and many broadleaved species, such as sycamore, ash and birch, also follow this pattern. Oak and beech, on the other hand tend to lay increment down fastest over a slightly longer period, namely between 25 and 50 years. Most of this increment, however, is unavailable for cutting except in the form of thinnings and so it is not so much the total increment that determines the amount of volume that can be removed annually without depleting the growing stock but rather the way in which that increment is distributed over the P Year classes.

In Great Britain at the present time the coniferous area, and consequently the coniferous volume, is heavily weighted towards the younger age classes and so the proportion of increment that can be removed in thinning and felling is considerably less than the total coniferous increment might suggest; the broadleaved position, on the other hand, is essentially the reverse. However, at the present time it is largely market forces that determine the amount of timber volume that is removed annually in Great Britain although controls to effect undercutting, or sanction of deliberate overcutting, may be introduced at times to meet specific national objectives. The discussion which follows relates only to gross annual increment.

Woodlands

The method used to calculate the increment of woodland crops on this occasion was to calculate from the Management Tables the appropriate increments per cent for a range of species and yield classes, i.e. expressing the current volume increment at a given age as a percentage of the standing volume per hectare at the same age. In the case of coniferous species the increments per cent were expressed as a percentage of the before thinning main crop volumes at ages which represented the mid-points of the P Year classes, e.g. at 15, 25, 35 etc. years. For broadleaves the percentages were based on the before thinning main crop volume plus the previous thinning in order to take account of the known underthinning of broadleaved crops in recent years.

One of the main advantages of using increments per cent is that it does not vary greatly with yield class. Consequently total standing volume has a greater influence on the increment total than the yield class itself and this enables the variations in standing volume which occur in various parts of the country, even in crops in the same species, age and yield class, to be taken into account.

The actual increments per cent used in the calculations are shown in Table 39. Other conifers and Mixed conifers, for which there are no yield models, were represented by Norway spruce Yield Classes 10 and 8 respectively whereas Other broadleaves and Mixed broadleaves were both represented by the category of sycamore, ash and birch Yield Class 4.

Application of these increments per cent to the appropriate standing volumes produced results which are shown in Table 40.

This table shows the gross current annual increment is 10.7 million cubic metres of which 8.1 million or 76 per cent is coniferous and 2.6 million or 24 per cent broadleaved. Of the coniferous increment Scotland holds 3.6 million or 44 per cent closely followed by England with 3.1 million or 38 per cent. The coniferous increment in Forestry Commission woodland exceeds that in private woodland in all three countries and is a reflection not only of the greater areas involved but also their somewhat older average age structure. On the other hand, over three-quarters of the total broadleaved increment occurs in England, and because of the Forestry Commission's small holding in the broadleaved re-

Table 39 Table of increments per cent

P year class	Age in 1980 (years)	Scots pine YC8	Corsican pine YC12	Lodgepole pine YC8	Sitka spruce YC14	Norway spruce YC12	European larch YC8	Jap./Hybrid larch YC8	Douglas fir YC14	Other conifer YC10	Mixed conifer YC8
1961–70	15	15.0	15.0	13.0	17.5	16.0	14.0	15.0	17.5	17.5	20.0
1951–60	25	10.8	9.0	9.4	11.6	12.9	9.4	9.7	10.8	14.6	17.1
1941–50	35	8.2	6.4	6.0	8.0	8.5	6.3	5.8	7.0	8.7	9.0
1931–40	45	6.2	4.6	4.2	5.1	5.8	4.2	3.8	4.7	5.9	6.3
1921–30	55	4.6	3.4	3.2	3.3	4.2	2.8	2.8	3.4	4.3	4.5
1911–20	65	3.4	2.4	2.6	2.2	3.1	1.9	2.1	2.5	3.2	3.3
1901–10	75	2.5	1.7	2.1	1.6	2.4	1.3	1.6	1.7	2.5	2.6
1861–1900	100	1.0	0.9	—	0.7	1.4	0.6	1.1	0.9	1.5	1.5
Pre 1861	150	0.3	0.3	—	0.3	0.6	0.1	0.3	0.3	0.7	0.7

P year class	Age in 1980 (years)	Oak YC4	Beech YC6	Sycamore YC6	Ash YC6	Birch YC4	Poplar YC6	Sweet chestnut YC6	Elm YC6	Other broadleaved YC10	Mixed broadleaved YC8
1961–70	15	17.5	17.5	13.0	13.0	17.5	15.0	17.5	17.5	15.0	15.0
1951–60	25	14.7	16.0	9.3	9.3	13.4	7.0	16.0	16.0	13.4	13.4
1941–50	35	6.7	6.9	5.2	5.2	5.7	3.5	6.9	6.9	5.7	5.7
1931–40	45	4.6	5.3	3.3	3.3	3.4	1.7	5.3	5.3	3.4	3.4
1921–30	55	3.8	4.3	2.3	2.3	2.3	0.9	4.3	4.3	2.3	2.3
1911–20	65	3.0	3.4	1.5	1.5	1.6	0.5	3.4	3.4	1.6	1.6
1901–10	75	2.4	2.8	1.1	1.1	1.1	0.3	2.8	2.8	1.1	1.1
1861–1900	100	1.5	1.7	0.6	0.6	0.6	—	1.7	1.7	0.6	0.6
Pre 1861	150	0.5	0.8	0.1	0.1	0.1	—	0.8	0.8	0.1	0.1

Table 40 Current annual volume increment in woodlands by category, ownership and country

Thousands of cubic metres overbark

Category	Forestry Commission woodlands				Private woodlands				Total woodlands			
	England	Wales	Scotland	Great Britain	England	Wales	Scotland	Great Britain	England	Wales	Scotland	Great Britain
Coniferous	1 688	1 111	2 203	5 002	1 436	290	1 409	3 135	3 124	1 401	3 612	8 137
Broadleaved	157	26	11	194	1 820	301	269	2 390	1 977	327	280	2 584
Total	1 845	1 137	2 214	5 196	3 256	591	1 678	5 525	5 101	1 728	3 892	10 721

Table 41   Current annual volume increment in woodlands by species group and P year class – Great Britain

Thousands of cubic metres overbark

Species group	P61–70	P51–60	P41–50	P31–40	P21–30	P11–20	P01–10	P1861–1900	Pre 1861	Total
Pines	378	747	337	341	330	57	45	35	4	2 274
Spruces	814	1 471	825	484	140	13	5	4	—	3 756
Larches	278	573	166	113	54	13	9	5	—	1 211
Other conifer	234	384	83	74	77	17	10	16	1	896
Total conifers	1 704	3 175	1 411	1 012	601	100	69	60	5	8 137
Oak	5	50	93	65	76	67	92	213	37	698
Beech	6	64	25	27	20	20	40	100	44	346
Sycamore, ash, birch	115	299	279	112	81	34	22	21	1	964
Other broadleaves	63	172	108	58	65	26	29	48	7	576
Total broadleaves	189	585	505	262	242	147	183	382	89	2 584
Total	1 893	3 760	1 916	1 274	843	247	252	442	94	10 721

source, its share of the broadleaved increment is less than 8 per cent of the England total.

The distribution of increment by P Year class and species group for all ownerships combined is shown in Table 41.

This table shows quite clearly the concentration of coniferous increment in the younger crops. No less than 21 per cent relates to crops under 15 years of age and 60 per cent to crops under 25 years. Most of the pine increment is Scots pine and this species accounts for virtually all the increment of this group in the older P Year classes. Sitka spruce accounts for over 70 per cent of the spruce increment and it is only in crops of P40 and older that Norway spruce forms a significant proportion of the total. Japanese/ Hybrid larch contributes nearly 80 per cent of the larch increment with European larch only of importance in the middle and older P Year classes. The major contributor to the Other conifer total is Douglas fir which accounts for over half of this category. The total coniferous increment is increasing substantially each year and as it does so the pattern of increment distribution between species alters with Sitka spruce becoming increasingly important, and the pines and larches less so.

With the exception of the P61–70 class, oak has fairly similar quantities in the P Year classes up to 80 years of age. However, no less than 36 per cent of the total oak increment is in crops over 80 years of age and this reflects the very substantial area and volume of this species present in these classes; beech shows a broadly similar pattern. However, syca-more, ash and birch, and indeed most of the other broadleaved species, show the increment to the present mainly in crops under 50 years of age; of these species and species groups birch is the most important accounting for nearly 17 per cent of the overall broadleaved increment total and its increment is particularly important in the P41–50 and P51–60 age groups.

Non-woodland trees

Rather less is known about the rates of increment of non-woodland trees and consequently a less detailed approach was adopted. In 1951 the increments per cent used were four per cent for volume in the $6\frac{1}{4}$–$9\frac{3}{4}$ inches Breast Height Quarter Girth class (20–31.5 cm dbh) and 1.2 per cent for volume in the larger size categories. In 1980 with the higher coniferous volumes and greater knowledge of the diameter distribution of broadleaved crops it was decided to produce increments per cent for each of the four size classes adopted. However, it was found that the distribution of volume by size classes in England, where the bulk of the volume and increment lie, were such that increment could be calculated quite satisfactorily using only two values namely 4 per cent for the volume in trees of 7–50 cm dbh and one per cent for the volume in trees over 50 cm dbh. These increments per cent were consequently applied to the standing volumes in each country and the following results obtained.

Table 42 Current annual volume increment by species group – non-woodland trees

Thousands of cubic metres overbark

Species group	Country			
	England	Wales	Scotland	Great Britain
Pines	23	5	8	36
Spruces	2	2	4	8
Larches	6	3	6	15
Other conifer	19	3	1	23
Total conifers	50	13	19	82
Oak	100	26	11	137
Beech	28	8	11	47
Sycamore, ash, birch	121	29	18	168
Other broadleaves	111	16	17	144
Total broadleaves	360	79	57	496
Total	410	92	76	578

From this table it will be evident that of the 578 000 cubic metres of gross increment in non-woodland trees England accounts for 410 000 cubic metres or 71 per cent, Wales for 92 000 or 16 per cent and Scotland for 76 000 or 13 per cent. These percentages are also generally applicable to the distribution of broadleaved increment by countries but with conifers England holds a rather smaller percentage (61 per cent) and Scotland a rather larger one (23 per cent).

Pines, especially Scots pine account for a substantial part of the coniferous increment but the increment in Other conifers in England at 19 000 cubic metres is also significant. Much of this total relates to cypresses present in our towns and cities and is consequently unlikely ever to be harvested. In broadleaves, oak is the major species with 137 000 cubic metres or 28 per cent of the broadleaved total followed by beech, sycamore and alder.

No less than 142 000 cubic metres, or over 25 per cent of the total of 578 000 cubic metres, is being added annually to the volume of trees of 50 cm dbh and over, although this is also the class which is experiencing the bulk of the fellings.

Part III

COMPARISONS WITH PAST SURVEYS

CHAPTER 17

Methods and results – woodlands

While the main purpose of any Census must be to produce results to the specifications set out in the objectives it is obviously desirable that these results should be capable of comparison with those of previous Surveys so that some measure of the changes that have taken place in the interim can be obtained. It would therefore seem sensible that such factors as definitions of woodland types, minimum woodland size, minimum diameter, timber length, etc., should remain unaltered. If this were the case the results would be compatible and, depending on their precision, the amount of change could be assessed and valid conclusions drawn. However, this procedure is not really practical for a number of reasons.

First, the objectives of surveys do differ. The type of questions that were being asked 15 years ago may not be appropriate to the present day problems and a change of objectives usually means that alterations need to be made to the data specifications.

Second, for many years the Imperial system was the accepted standard for units of measurement in forestry and indeed, in most other industries in Great Britain, and so land areas were traditionally expressed in acres, diameters in inches, tree girths in inches quarter girth and timber volumes in cubic feet quarter girth (Hoppus Feet). In 1971 the timber industry moved to the metric system and adopted units appropriate to it. Inevitably, however, the new units did not equate exactly with their Imperial counterparts and this involved awkward conversion factors. Consequently features such as minimum diameters, lengths, etc., whilst usually close to their previous values, now differed and meant that although the differences might be minor direct comparison with quantities in the past was no longer possible.

It must be stressed that when changes are made to survey specifications they are only instituted after a great deal of thought. The need to compare results with those of the past is always at the forefront of the thoughts of the survey designer and when changes are made it is because the need for change at the present time is greater than the need for direct comparison with the past. Very often the changes are so small as to have very little effect on the comparison; others, of course, have effects which cannot always be directly quantified. Some of the major difficulties in comparing the results of the present survey with those of the past are shown below together with a note on how they were overcome. It will be appreciated that on occasion there were a number of options available and each one would have produced a different solution. Where this was the case the results were often calculated in more than one manner and the most reasonable solution accepted.

Difference in minimum wood size

The minimum wood size has differed in almost every Census and the actual values used are quoted in Chapter 1 together with an explanation for the choice where this was known. Having adopted 0.25 of a hectare in 1980 the problem was then to compare the results with those of the 1947 Census which adopted a minimum wood size of five acres (2.02 hectares) and 1965 where the minimum was one acre (0.40 of a hectare).

There was a survey of small woods in 1951 which complemented the results of the 1947 Census and provided an estimate of the area of woods between one and five acres. It was based on two independent samples of six inch to one mile (1:10560) maps distributed throughout the country and the area of any woods which lay within the prescribed limits were measured on them. There was no question of confusing these woods with woods of five acres and over because the maps used were the survey maps used in the main survey and consequently all these larger sized woods had already been visited and the boundaries demarcated. Sample woods were then visited on the ground to enable the total area to be allocated to forest types. Because of the relatively small sample size it was not practical to produce estimates on a county basis and only the country totals were calculated and published.

To relate the results of this Survey to those of 1980 it is necessary to make a further addition to the 1951 total in order to account for the reduction in the minimum wood size from one acre (0.4 of a hectare) to 0.25 of a hectare. Obviously since one acre was the minimum size assessed in 1951 there was no information on woods smaller than this from the survey and it was necessary to use information gained in the 1980 Census. From this survey the number and area of woods that lay between 0.25 and 2.0 hectares were known, as these limits were deliberately chosen to aid any comparison with the 1947 Census. However, by the time of first publication of the 1980 Census results the data had not been analysed to show the distribution of these woodland blocks over the size range and so it was assumed that the total area of woods between 0.25 and 2.0 hectares could be subdivided into 35 equal parts, each representing 0.05 of a hectare. Consequently the area between 0.25 and 0.4 of a hectare could then be assumed to be 3/35 of the total. This area when added to the 1951 total for each country gave an estimate of the total woodland area between 0.25 and 2.0 hectares as it was in 1951. This total in turn was added to the area of woods of five acres and over in 1947 to provide an estimate to the same base as the 1980 Survey. A similar approach was used to adjust the area in 1965 to cater for the then minimum wood size of 0.4 of a hectare.

If the total areas of small woods found in the 1951 and 1980 Surveys had been reasonably comparable in area the procedure adopted for calculating the area of woodland between 0.25 and 0.4 of a hectare would have been considered fair and reasonable. However, after the adjustments had been made the results of the two Surveys were compared on a country basis and substantial differences in the area of these small woods were found. In England the difference between the two estimates of woodland between 0.25 and 2.0 hectares was about 22 000 hectares, an increase of 45 per cent on the 1951 total, in Wales about 1 800 hectares, a rise of 15 per cent, and in Scotland about 4 500 hectares or a rise of 26 per cent. As the overall adjustment for woods between 0.25 and 0.4 of a hectare for Great Britain only amounted to some 9 000 hectares it was clear that other factors must be involved. Possible reasons for the increases were:

    a. the 1951 estimate was based on a relatively small sample of maps, many of which were well out of date;

    b. some fragmentation of larger blocks was likely to have occurred as a consequence of the increased pace of house building, road building, power lines, etc.;

    c. where there is colonisation it usually occurs as a series of small blocks which may coalesce later to form larger woods. These areas of colonisation would normally post-date the maps used in the 1951 Survey but even where they existed in 1951, and were of minimum acceptable size, they were unlikely to have been picked up during the course of the field work;

    d. new planting of small blocks particularly along motorways, trunk roads and in areas of urban development or redevelopment had taken place.

All four factors are likely to have had some effect but fragmentation and new planting seem likely to account for the largest part of the discrepancy between the two sets of estimates.

Changes in definition

Generally speaking the definitions of the main forest types do not alter greatly from one census to another but it is inevitable that particular forest types will be introduced into a Census to answer specific questions or to subdivide categories that were too diverse. For example, in 1924 a category of 'Uneconomic' was introduced to cover woods which were not maintained for timber production but primarily served some other purpose and was largely composed of amenity woods, shelterbelts and park timber. A substantial part of this area was undoubtedly what later would have been termed High Forest but nothing was known about it except as a total as it was not subdivided by category, species or age. Consequently it is impossible to compare the High Forest total in subsequent surveys with that in 1924 because the area of this category was understated. Similarly in 1947 the 'Devastated' category was introduced to cover crops from which most of the merchantable trees had been removed, and the remnants incapable of satisfactory development on their own. This category was unique to the 1947 and 1951 Surveys and mainly catered for crops that originally had been in one of the productive types of High Forest or Coppice but were then currently not classifiable as Scrub or Felled. The remnants often gave little clue as to whether the previous crop had been predominantly coniferous or broadleaved and this made subsequent assessment of the area and volume losses of the various forest types extremely difficult. Bearing in mind the very extensive war-time fellings comparison with past results were somewhat academic anyway but it also created problems in comparing the results with those of subsequent surveys.

In 1965 a classification had to be introduced to meet the assessment needs of crops, particularly broadleaved crops, that were then in a transition stage. For example many crops that had been felled had coppiced and produced multiple stems but the likelihood of these crops ever being worked as coppice was remote. Similarly colonisation had occurred on felled sites and wasteground, often forming dense and impenetrable thickets. There were also problems with coppice crops that were well beyond the age of normal coppicing and were in some cases more akin to High Forest and classified as such, whilst others were virtually derelict and the crops unworkable and unsaleable. There was still evidence of exploitation in many of the stands with remnants of the previous crop still in existence. Consequently it was found necessary to subdivide the scrub category into Utilisable and Unutilisable Scrub to cater for the wide variety of crop characteristics present.

In 1980 some changes were made but they tended to be relatively minor. For example, a sub-category of Broadleaved High Forest was introduced termed Broadleaved High Forest of Coppice Origin to allow for crops which still had the characteristics of coppice but which were now well beyond the age of normal coppice working and would be expected to be singled in future to form normal High Forest. The categories Mainly Coniferous and Mainly Broadleaved High Forest were accepted as the major subdivision of the High Forest category to accord with international convention.

Consequently it will be seen that no two Surveys have had the same definitions and while this is unfortunate the changes were made to answer specific questions relating to crop conditions at the time and it is almost inevitable that similar changes will be made in the future.

An added difficulty is that crops, especially broadleaved crops do not necessarily remain in one forest type for the whole of the rotation. It is possible for coppice crops to be grown on to form High Forest, High Forest can be coppiced, heavy selective felling can lead to a crop being downgraded to Scrub, Scrub crops in turn can develop in a manner that enables them to be re-classified into one of the productive categories. Direct comparison of the area of a forest type at one survey with the same category at an earlier date must therefore be made with caution and it is perhaps best to consider crops to be either coniferous or broadleaved and then to make the comparisons on this basis. In this way the movement of crops between categories is only of importance in the case of Mixed High Forest. When comparisons have therefore been made to establish whether the area of broadleaved crops has increased or decreased it has been done on the basis of combining the areas of all the appropriate High Forest, Coppice, and Coppice with Standard and Scrub crops as found in the 1947 and 1965 Censuses and comparing them with exactly the same combination of categories in the 1980 Census. Devastated crops have been omitted from the 1947 and 1951 totals in making the comparisons as such crops had been classified on the basis of the remnants after felling and so were not necessarily representative of the crops that had been there beforehand.

Changes in boundaries

As the national boundaries are unchanging there is usually no difficulty, once the problem of minimum woodland size has been overcome, in comparing the total woodland areas of the individual countries for two or more Surveys. Comparing results for regional areas within countries can, however, create major difficulties if boundary adjustments have taken place on a big scale. There was a major reorganisation of county boundaries in England and Wales in 1974 and in Scotland in 1975; in some cases the changes were minimal, in others whole new counties were established with boundaries that had little in common with those that had previously existed. Under these circumstances comparison of the results of woodland surveys for individual counties becomes at best difficult and at worst virtually impossible. The only practical solution would have been to reassemble the data for one of the Surveys to match the revised county boundaries of the other but this would have constituted an enormous task and was not contemplated seriously. What was done in the case of those counties in England and Wales where there had been virtually no change in the boundaries was to bring the two surveys to a common base by excluding the area of woods of 0.25–2.0 hectares from the 1980 results and then comparing the total with that found in 1947. Where there had been changes in the boundaries of the counties between the two surveys further adjustment was made to cater for the change in land area, usually on a pro rata basis.

The adoption of these procedures raised two reservations. First, the assumption that the area of woods between 0.25–2.0 hectares found in the 1980 Survey does not contain any blocks which have arisen through fragmentation of woods that were classed as being of two hectares and over in the earlier survey. Second, the use of a pro rata adjustment for land area alterations presupposes that the woodland is spread evenly over the county and that a reduction of 5 per cent in the county land area will in turn result in a 5 per cent reduction in woodland area. In some instances this is manifestly untrue but without a great deal of work there was no real practical alternative to the approach adopted. In Scotland, where the sampling was based on Conservancies, the same type of difficulties arose in that Forestry Commission boundaries alter from time to time. Consequently, any comparison of the results on a Conservancy basis, whether it be in Scotland, England or Wales, created the same type of problem and the solution was the same as that adopted for individual counties.

Despite the many difficulties which exist in making comparisons of the results of the 1980 Census with those of the past an attempt has been made to reconstruct the main totals of the 1947 and 1951 Censuses and the 1965 Census to a common minimum wood size of 0.25 of a hectare.

Total area

An assessment of total areas at the three survey dates is shown in Table 43 below.

Table 43 Estimated area of woodland 0.25 ha and over by countries

Thousands of hectares

Year	England Area	% of land area	Wales Area	% of land area	Scotland Area	% of land area	Great Britain Area	% of land area
1947	805	6.2	141	6.8	530	7.2	1 476	6.7
1965	892	6.8	202	9.7	657	9.0	1 751	7.8
1980	948	7.3	241	11.6	920	12.6	2 108	9.4

Note: Figures may not always add to totals due to rounding errors.

This Table shows that the woodland area of Great Britain increased by some 275 000 hectares between 1947 and 1965 and by a further 357 000 hectares between 1965 and 1980. The major increase in terms of both actual and percentage values has been in Scotland where the total rose over the whole period by some 390 000 hectares or 74 per cent. England during the same period showed an increase of some 143 000 hectares or 18 per cent and Wales 100 000 hectares or 71 per cent.

Woodland ownership

Of the total of 2 108 000 hectares in 1980, 892 000 hectares were in Forestry Commission ownership and 1 216 000 hectares in the hands of private owners. The relative positions at the three survey dates are shown in Table 44. Here it has been assumed that the total area of woodland in blocks between 0.25 and 2.0 hectares in the 1947 and 1951 Censuses and between 0.25 and 0.4 of a hectare in 1965 were privately owned. As the areas in each country are relatively small this assumption results in only minor differences to the percentages that were calculated using the unadjusted Census results. The percentages appropriate to the common woodland base totals are shown in Table 45.

Table 44 Relation between Forestry Commission and private woodlands at three survey dates

Minimum woodland area 0.25 ha   Thousands of hectares

Country	Forestry Commission			Private woodlands			Total		
	1947	1965	1980	1947	1965	1980	1947	1965	1980
England	117	234	255	688	658	693	805	892	948
Wales	37	117	139	104	85	102	141	202	241
Scotland	98	304	498	432	353	422	530	657	920
Great Britain	252	655	892	1 224	1 096	1 216	1 476	1 751	2 108

Note: Figures may not always add to totals due to rounding errors.

Table 45 Relation between Forestry Commission and private woodlands at three survey dates

Minimum woodland area 0.25 ha   Per cent

Country	Forestry Commission			Private woodlands			Total		
	1947	1965	1980	1947	1965	1980	1947	1965	1980
England	8	13	12	47	38	33	55	51	45
Wales	3	7	6	7	5	5	9	12	11
Scotland	7	17	24	29	20	20	36	37	44
	17	37	42	83	63	58	100	100	100

Note: Figures may not always add to totals due to rounding errors.

Table 44 shows the substantial increase in the area of woodland owned by the Forestry Commission in all three countries over the whole period and also the reduction in the private woodland area between 1947 and 1965 largely as a result of the acquisition of derelict and felled woodland sites for restocking by the Forestry Commission; a considerable area of felled woodland was also converted to agricultural use, particularly grazing. Since 1965, however, there has been an increase in the privately owned area, largely as a result of afforestation, which has brought the private totals back to near their 1947 levels. When the positions of individual countries are examined it will be seen that between 1947 and 1980 the Forestry Commission area in England more than doubled, in Wales it nearly quadrupled and in Scotland there was a five-fold increase. However, in both England and Wales most of the increase had been completed by 1965 whereas in Scotland the increase has been equally divided between the two periods. Looked at in percentage terms England's share of the total woodland area has declined from 54 per cent in 1947 to 45 per cent in 1980 whilst Wales has shown a rise of 2 per cent and Scotland one of 7 per cent over the same period.

Distribution by forest type

Because of changes in the crop classification in the various surveys there are problems in making direct comparisons of the areas of individual forest types. The best estimate of the areas at each of the three survey dates is set out in Table 46 but it must be realised that the further the results are subdivided the greater the effect the assumptions are likely to have on the individual categories. The results should therefore be used with caution.

Mainly coniferous high forest

The area of woodland under conifers has risen from some 397 000 hectares in 1947 to 922 000 hectares in 1965 and to 1 317 000 hectares in 1980; the proportion of woodland area the forest type represents has risen from 27 per cent to 62 per cent. The rate of increase was most marked during the period between 1947 and 1965 when the coniferous area increased by more than half a million hectares. The relative importance of the major coniferous species at the time of the three surveys is shown below with the sudden appearance of Lodgepole pine in 3rd position in 1980 noteworthy.

Mainly broadleaved high forest

Both the area of Mainly Broadleaved High Forest and its proportion were reduced between 1947 and

Table 46 Forest type distribution at three survey dates

Minimum woodland area 0.25 ha   Thousand hectares

Forest type	1947		1965		1980	
	Area	%	Area	%	Area	%
Mainly coniferous high forest	397	27	922	53	1 317	62
Mainly broadleaved high forest	380	26	352	20	564	27
Total high forest	777	53	1 274	73	1 881	89
Coppice with standards	95	7	11	1	12	1
Coppice	50	3	19	1	27	1
Total coppice	145	10	30	2	39	2
Scrub	213	14	373	21	148	7
*Cleared	341	23	74	4	40	2
Total	1 476	100	1 751	100	2 108	100

* Includes devastated in 1947.

Table 47 Relative importance of coniferous species at three survey dates

Year	1st	2nd	3rd	4th
1947	Scots pine	Sitka spruce	Norway spruce	European larch
1965	Scots pine	Sitka spruce	Norway spruce	Jap./Hybrid larch
1980	Sitka spruce	Scots pine	Lodgepole pine	Norway spruce

1965, but since that time both parameters have increased to such an extent that there is now substantially more Mainly Broadleaved High Forest than there was in 1947. The initial reduction came about as woodlands were acquired, cleared and replanted, often with conifers, as part of the post-war policy of returning the woodlands of Great Britain to full production. The increase since 1965 is a combination of several factors; land has been planted with broadleaves; there has been colonisation by birch, sycamore and ash, leading to High Forest; and most noticeable, the development of Scrub either naturally, or through deliberate management into a form capable of being reclassified as High Forest. The relative importance of the major broadleaved species in High Forest at the time of each Survey is shown below in Table 48 and shows a virtually unchanged pattern.

Table 48 Relative importance of broadleaved species at three survey dates

Year	1st	2nd	3rd	4th
1947	Oak	Beech	Ash	Birch
1965	Oak	Beech	Ash	Sycamore
1980	Oak	Beech	Ash	Birch

Coppice and coppice with standards

There has been a very substantial drop in the area of Coppice with Standards and Coppice since 1947. The area then classed as Coppice types amounted to 145000 hectares although the proportion actually being worked under these systems was unknown. The area had declined to about 30000 hectares in 1965 and has since risen to about 39000 hectares most of which appears to be worked. The Counties of Kent and East and West Sussex currently account for nearly 70 per cent of the area of coppice system crops, although there is evidence elsewhere of a renewed interest in this form of management. However, except for such isolated examples, there is very little fresh involvement in working either Coppice or Coppice with Standards. Stands previously classed as being under one or other of the Coppice systems have been cleared and restocked or have been allowed to develop naturally, some into Scrub, others into Broadleaved High Forest.

The principal species of Coppice currently are Sweet chestnut followed by hornbeam and hazel. In 1947 hazel was clearly the dominant coppice species but oak was then, and still remains, the most widely used species of standard.

Scrub

The area of Scrub excluding Devastated has fallen over the years from about 213000 hectares, or 15 per cent of the woodland area in 1947, to 7 per cent in 1980. There was a rise in 1965 largely because of the classification adopted. As might be expected, Scrub plays a minor part in Forestry Commission and Dedicated and Approved woodlands, and in fact 87 per cent lies in 'Other' private ownership. Birch has been, and still is, the most widespread species followed by oak and hazel.

Cleared

The area classed as Cleared has shown a very dramatic fall since 1947 but the extent of this reduction must be treated with some caution. In 1947 a considerable number of the map editions used for the Census were out of date and areas were shown as woodland which in some instances had been bare of trees for many years. Although at the time of inspection the majority of these sites were being grazed, evidence in the form of the stumps of the previous crop was still present and unless there was a clear indication that the sites were being permanently converted to agriculture by stump removal, etc., they were still considered to be woodland and were classed as Felled prior to September 1939. In the last 30 years many of these sites have become restocked with trees but others have been totally converted to agricultural use or are currently shown on maps as rough grazing although evidence of stumps can still be found; in neither case are they now considered as woodland. Most of the decrease of 300000 hectares in the area of Cleared is therefore a true reduction but part must be attributed to a change in land use or in classification.

Overall broadleaved position

As noted previously it is difficult to make direct comparisons with past results, especially where broadleaved species are concerned, but it is possible to draw some conclusions at national level if the areas of certain forest types are combined. If the broadleaved areas of High Forest, Coppice with Standards, Coppice and Scrub are totalled for each Survey and allowances made for the effects of small woodland blocks, it would appear that the total area in England and Wales is probably slightly more than it was 30 years ago, whereas in Scotland it is less. On balance the total area of broadleaved woodland is much the same as it was in 1947. However, its composition has certainly changed and there are now about 70000 hectares less of oak, but more sycamore, ash and birch than in 1947. When making the comparison with previous Surveys it needs to be remembered that there is an element of coniferous scrub for which allowance needs to be made.

Standing volume

It is estimated that in 1980 there were 197 million cubic metres overbark of timber in the woodlands of Great Britain, comprising 106 million cubic metres of coniferous timber and 91 million cubic metres of broadleaved timber. The Forestry Commission accounted for 32 per cent of the timber stocks, private owners for 68 per cent.

Volume estimates were produced for both the 1947 and the 1965 Surveys, and whilst the standards of measurement were similar to those of 1980 the overall estimates are affected by the minimum area and classification differences.

The 1947 volume of 108 million cubic metres (including an allowance for the volume in small woods) rose to 124 million cubic metres in 1965 (allowing for the volume in woods of 0.25–0.4 of a hectare) and has since risen to 197 million cubic metres in 1980.

The rate of volume increase which was relatively small between 1947 and 1965, and has been much larger since then, can be expected to continue to rise rapidly as the large areas planted during the last 25 years move into the measurable size category. Most of this volume increase will be of coniferous timber.

The percentage of volume found by ownership and by species are shown in the Tables 49 and 50.

Table 49 Volume by ownership classes as a percentage of total standing volume

Ownership class	Category	1947	1965	1980
Forestry Commission	Coniferous	11	22	30
	Broadleaved	4	3	2
	Total	15	25	32
Private	Coniferous	31	29	24
	Broadleaved	54	46	44
	Total	85	75	68
All	Coniferous	42	51	54
	Broadleaved	58	49	46
	Total	100	100	100

There has thus been a substantial change in the composition of the growing stock from broadleaved to coniferous timber during the last 30 years. However, the proportional change was greatest between 1947 and 1965 when 9 per cent of the volume moved from broadleaves to conifers, but since that time only a further 3 per cent has changed categories. This is a reflection of the increased area of woodland carrying volume that is present in 1980.

The Forestry Commission share of the volume has more than doubled to 32 per cent in 1980 but even so, private owners still hold 68 per cent of the total standing volume.

Table 50 Distribution of volume by major species groups in high forest as a percentage of total standing volume

Species group	1947	1965	1980
Pines	20	23	18
Spruces	8	14	21
Larches	8	10	9
Oak	27	20	17
Beech	13	10	8
Sycamore, ash and birch	10	10	12

Within the broad context of increased standing volume in both coniferous and broadleaved timber, there is a developing trend in distribution as newly planted areas begin to produce measurable timber and the older, traditional species consequently become less important in percentage terms. The proportion of volume in larches is virtually unchanged; pines, after a surge in 1965 are only slightly less than their 1947 levels, whilst the spruces have moved from an 8 per cent share in 1947 to 21 per cent in 1980.

CHAPTER 18

Methods and results – non-woodland trees

The problem of the minimum size of wood adopted in the various Census surveys affects not only woodland comparisons but also those of non-woodland tree surveys. This is because in 1951 any trees in woods or copses under one acre (0.4 of a hectare) in extent or 22 yards (20.1 metres) in width were included with the isolated trees in the Hedgerow and Park Tree Survey whereas in 1980 non-woodland trees comprised individual trees or groups which in general were under 0.25 of a hectare in area or under 20 metres in width. Therefore when comparisons are made between the results of the two Surveys the 1951 totals need to be reduced to allow for the problem of the minimum wood size. In the case of tree numbers a method similar to the calculation process for woodland was used, i.e. determine the area of woods from the 1980 Census that lay between 0.25 and 2.0 hectares in each country, take 3/35 of the total and multiply this figure by the average number of trees found per hectare in Linear Features in that country. The values actually used in the calculation were England 290, Wales 375, Scotland 424 trees per hectare. Again it has to be assumed that the area of small woods is spread evenly throughout the size range and also that Linear Features best reflect the average representation of numbers for these small woods or copses. One might argue that using the average number of trees for High Forest or Clumps might be a more appropriate average to use but it was felt that as these small blocks tend to have somewhat variable stocking the High Forest average would be rather too high and that Linear Feature values would be more appropriate than the Clump value because there was likely to be a higher proportion of Linear Features in the 0.25–0.4 of a hectare range than Clumps which could be composed of as few as two trees.

Another major difference involved the diameter limits adopted at the two surveys. In 1951 all measurements were in Imperial units. Saplings were defined as being trees that were straight and of good form and were between 3 and 6 inches BHQG (9.7–19.4 cm dbh). Timber trees had to have a minimum BHQG of 6 inches (19.4 dbh) and at least 10 feet (3.05 metres) of straight usable timber.

In addition there were trees classed as 'Shorts' which were trees of 6 inches BHQG and over with between 6 and 10 feet (1.83–3.05 metres) of timber, and 'Firewood' trees which were of all girths 3 inches BHQG and over where the bole length was less than 6 feet, or the trees were affected by rot or decay.

The 1980 tree classification counted and measured trees once they had attained a diameter of 7 cm (2½ inches Quarter Girth) at breast height (1.3 metres) provided they had a straight persistent axis and were not pruned in such a way as to restrict growth.

Trees grown as pollards were accepted but much of the pruning done in towns and cities in the name of pollarding resulted in the trees being omitted from the tree count altogether. Trees under 7 cm dbh were counted and included in the survey provided they satisfied certain criteria in that they had to be at least 1.5 metres tall, have a persistent axis, and were individuals not coppice shoots. These are obviously potential trees for the future and of a size whereby they have a reasonable chance of survival. Exceptions were made to the minimum height criterion where the trees were planted as windbreaks or as roadside planting where they were seen to be protected by fencing or tree guards and again consequently there was a reasonable likelihood of their reaching maturity. The reason for the much more detailed approach in the 1980 Survey, particularly in the case of trees below the measurable limit, was that there was obviously a great deal of concern about the ability of the non-woodland tree population to maintain itself and unless there is a reasonable estimate of tree numbers in all the classes from the very smallest to the largest there can be no clear indication as to the prospects for the future. The inclusion of the under 7 cm class does not affect the comparison with the 1951 Census as it is an entirely new class but the 1980 Census, of course, started counting and measuring trees at a rather smaller diameter than was the case in 1951. In 1951 the smallest tree classed as measurable in the woodland survey was 2½ inches BHQG (8.1 cm) but the minimum adopted for the Hedgerow and Park Timber Survey was 3 inches (9.7 cm) because that was the limit adopted for the 1938 survey. This highlights the difficulty of selecting a limit such as a minimum diameter, and trying to continue to use it over a long period. There comes a time when comparability with other measurements then being taken is more important than comparability with the past.

The problem was therefore to look at means whereby the 1951 tree number total could be adjusted to include the likely tree numbers that would lie between 10 cm (3 inches BHQG) and 7 cm. If one assumed that tree numbers are spread evenly throughout the girth range then extending the limits of the class would have resulted in an increase in the 1951 numbers of saplings of about one quarter. However, as coppice stems of requisite girth and straightness were included in 1951, but generally not so in 1980, the total would need to be reduced but by an unknown factor. It seemed simplest in this case to make no adjustment to the totals as found in 1951 and 1981 but rather to look at the extent of the differences in the final results and see if the differing size limits could account for them. The difference between the tree numbers in the 3–6 inch BHQG class in 1951 and those in the 7–20 cm dbh class in 1980 appear to be much too great to be solely a problem of differing class limits.

Therefore some comparison of total tree numbers is possible but not by size classes. This is partly because the class limits are not compatible as a result of the introduction of the metric system but mainly because the diameter distributions in 1951 covered only saplings and timber trees; the numbers of Short trees and Firewood trees were not summarised by size class and the basic information, in any case, was not always recorded. The 1980 base, however, is now a better one for the future. All trees found, provided they met the definition of the class, were counted and recorded in their appropriate diameter class as individuals and any subsequent classification, based on their timber qualities, was secondary.

When comparison of non-woodland tree volume is considered, adjustments were again needed to effect this. Volume had first to be imputed to the 7–20 cm dbh class, which was not measured for volume in the 1951 Survey, and this was then added to the 1951 volume total. The volume arising from this source is obviously of minor consequence because of the small size of the trees, and it was considered sensible merely to add the volume found in the 1980 Census in this class, which amounts to less than 4 per cent of the 1980 total, to the 1951 total. However, a further adjustment needs to be made to the 1951 volume because it contains volume in woods between 0.25 and 0.4 of a hectare. This has to be deducted from the 1951 total and was calculated by determining the area involved, i.e. 3/35 of the area in woods of 0.25–2.0 hectares and multiplying

this area by the mean volume of Broadleaved High Forest in woodlands as information on volumes per hectare were not available for either Clumps or Linear Features. The two calculations worked in opposite directions and usually effectively cancelled each other out so that the effects of the adjustment were minimal.

Two other factors have to be borne in mind although they do not affect the comparison directly. First, the levels of sampling adopted in each of the two surveys varied considerably. In 1951 the ground area sampled was approximately one hectare in every 8 000 and was intended to supply information at the Conservancy and country level whereas in 1980 the ground area sampled was one hectare in every 900 hectares in England and Wales and every 2 500 hectares in Scotland and aimed to provide estimates, although obviously of fairly low precision, for counties and Regions. Thus the 1980 Survey was much more intensive than the 1951 Survey; it was also based on soil stratification which the earlier one was not and there are consequently fewer grounds for doubt as to the accuracy of the 1980 results. When the non-woodland survey was carried out in 1965 at the same intensity of sampling as in 1951 a substantial change in volume appeared to have taken place in South East England Conservancy, a finding which has not been borne out by the 1980 results. On the other hand the 1980 Survey showed a major change to have taken place in East Scotland Conservancy compared with the 1951 Survey (Scotland was not sampled in 1965) but there is no way of establishing how much of the difference is due to real change and how much is attributable to the relatively low sampling levels adopted in the earlier survey.

Another factor which makes comparisons difficult with the past is the very substantial loss of elm. In 1951 elm accounted for one fifth of the non-woodland tree volume in this country; it was a tree that was widespread in its distribution and in some cases was the predominant tree. Elm has now dropped from being second in importance in the country in volume terms to eighth and Dutch elm disease is expected to continue to take its toll. Currently the results of the two Surveys can only be compared on a country basis so the extent of loss on a local scale cannot be measured. The higher intensity of measurement in the 1980 Survey should however provide a better basis for measuring the degree of change in the future.

All the factors discussed so far make it difficult for detailed comparison of the results of one survey to be made with another but it is possible to draw some general conclusions.

Comparison of tree numbers

In 1951 the total number of trees 7 cm dbh and over

recorded for Great Britain was 73.3 million. This figure needs to be reduced to allow for trees in woods of 0.25–0.4 of a hectare to make it comparable with the 1980 total. The adjusted 1951 total is estimated to be 70.3 million trees. During the last 30 years, therefore, the total number of non-woodland trees has apparently risen from about 70 million to 88 million trees, a rise of 18 million or 25 per cent. This is a substantial increase by any standard, especially when one considers the loss in tree numbers as a consequence of Dutch elm disease. However over the period there undoubtedly has been large scale planting of motorway verges for the Department of Transport, and along main roads and in public open spaces by local authorities; there has also been a major increase in numbers in urban areas, particularly of garden and street trees. Many of these trees are now of measurable size and can thus substantially affect numbers in the 7–20 cm dbh class. Changed standards of inclusion may also have altered both total tree numbers and the pattern of size distribution to some degree. However, since 1951 it seems likely that there has been a near doubling of tree numbers in the 7–20 cm dbh class, and an increase of over 40 per cent in tree numbers in the 31–50 cm dbh size class. This situation is apparent in each of the three countries but it must be stressed that there are regional and local variations in the pattern and the recruitment position will not be as satisfactory in some areas as in others.

Volume comparison

The results of the 1951 and 1980 Surveys are shown in the following Table after adjustments to the 1951 figures to exclude the volume occurring in woods of 0.25–0.4 of a hectare and to include an allowance for the fact that volume measurements in 1951 were confined to trees of over 20 cm dbh.

Table 51   Standing volume in 1951, after adjustment, and in 1980

Millions of cubic metres

	1951	1980
Coniferous volume	2.1	2.9
Broadleaved volume	26.2	25.0
Total	28.3	27.9

These figures indicate that there has been a rise in coniferous volume in the 30 years between the two Surveys. Increases have taken place in both England and Scotland whilst the volume in Wales has remained constant. In the case of broadleaved volume there has been a slight reduction in the Great Britain total, with Wales and Scotland showing increases and England a substantial drop mainly due to the loss of elm. The change in the individual species

Table 52   Distribution of standing volume by species in 1951 and 1980   Millions of cubic metres

1951			1980		
Species	Volume	Per cent	Species group	Volume	Per cent
Scots pine	0.8	3	Pines	1.2	4
Norway spruce	0.2	<1	Spruces	0.2	1
European larch	0.2	<1	Larches	0.4	1
Other conifers	0.9	3	Other conifers	1.1	4
Total conifers	2.1	7	Total conifers	2.9	10
Oak	8.7	31	Oak	8.8	32
Beech	2.4	9	Beech	2.7	10
Sycamore	2.1	7	Sycamore	2.6	9
Ash	3.7	13	Ash	4.1	15
Birch	0.1	<1	Birch	0.4	1
Sweet chestnut	0.1	<1	Sweet chestnut	0.4	1
Elm	5.5	20	Elm	1.0	4
Other broadleaves	3.6	13	Other broadleaves	5.0	18
Total broadleaves	26.2	93	Total broadleaves	25.0	90
Total	28.3	100	Total	27.9	100

contributions for Great Britain is set out in Table 52; the figures for 1951 have been adjusted and the additional volume added to species on a pro rata basis.

The coniferous species groups recognised at each Survey do not correspond exactly but it appears likely that pines, larches and Other conifers are rather more important now than they were in 1951, whilst spruces have maintained their previous level.

Among the broadleaves, the significant change is the position of elm which has dropped from being the second most important species overall, to eighth in this abbreviated list. The volumes and percentages of the more important named species have remained relatively constant with sycamore showing a modest increase and Other broadleaves a substantial increase over the 1951 levels.

Part IV

CONCLUSIONS

CHAPTER 19

Conclusions

The main conclusions from the 1979–82 Census of Woodlands, which had an operative date of 31 March 1980, were as follows:

1. The total area of woodland in Great Britain occurring in woods of 0.25 of a hectare and over in extent was 2 108 397 hectares representing 9.4 per cent of the land and inland water area of Great Britain.

2. Private woodlands, including those in the Dedication and Approved Woodland Schemes, in individual private ownership, in the hands of Water Authorities, the Ministry of Defence, etc., account for 1 216 700 hectares or 58 per cent of the total and Forestry Commission woodlands for 891 700 hectares or 42 per cent of the total.

3. England holds 947 700 hectares of which private woodlands account for 73 per cent, Wales holds 240 800 hectares of which 42 per cent is privately owned and Scotland 919 900 hectares of which 46 per cent is private.

4. Productive woodland, i.e. High Forest, Coppice with Standards and Coppice amount to 1 920 400 hectares or 91 per cent of the total; 1 045 900 hectares or 54 per cent are privately owned and the balance of 874 500 or 46 per cent is managed by the Forestry Commission.

5. High Forest occupies an area of 1 881 200 hectares of which Coniferous High Forest accounts for 1 316 800 hectares of which 819 300 hectares or 62 per cent is owned by the Forestry Commission. Broadleaved High Forest accounts for 564 400 hectares of which private owners hold 510 300 hectares or 90 per cent.

6. The age class structure of Coniferous High Forest is heavily weighted towards the younger age classes with over one million hectares, or 76 per cent of the total, having been planted since the war. This percentage is reasonably similar in both private and Forestry Commission wood-lands.

7. As a result of felling, and the acceptance of younger crops as High Forest which were previously classed as Coppice or Scrub, the broad-leaved woodland now has a more balanced age distribution. However, there is a relative shortage of crops in the P71–80 age class and a relative surplus in the decades either side of 1950. Over 35 per cent of the Forestry Commission broadleaved area has been planted since 1950 whereas the proportion in private ownership in the same P Year classes is only 20 per cent.

8. Coppice types amount to 39 100 hectares or 2 per cent of the woodland total of which Coppice with Standards account for 11 600 hectares and Coppice for 27 500 hectares. Almost all the area is privately owned.

9. Scrub woodland accounts for 148 200 hectares or 7 per cent of the woodland total and is again almost wholly in private ownership.

10. Cleared woodland totals 39 800 hectares or about 2 per cent of the woodland total. Approximately 30 per cent is held by the Forestry Commission and the balance by private owners. Most of this is likely to be replanted within the next few years.

11. The total standing volume in High Forest and Standards over Coppice is 197.4 million cubic metres overbark of which 106.3 million cubic metres is coniferous and 91.1 million cubic metres broadleaved. Private woodlands hold only 44 per cent of the coniferous volume but over 95 per cent of the broadleaved volume.

12. The volume of timber in non-woodland trees amounts to 28 million cubic metres of which 75 per cent is in England, 14 per cent in Wales and 11 per cent in Scotland. Eighty-nine per cent of the volume in this category is broadleaved.

Appendices

1. Glossary of terms

2. List of species recorded and reported upon – Woodlands
   List of species recorded but not reported upon individually – Woodlands
   List of species recorded and reported upon – Non-woodland trees
   List of shrub layer species recorded

3. Woodland type distribution by county, region and country

4. Woodland density by county, region and country

5. Predominant coniferous and broadleaved species in High Forest by county, region and country

6. Description of soil strata

7. Distribution of soil groups by county in England and Wales and by Conservancy in Scotland

8. Non-woodland tree information by county, region and country

9. Criteria for the assessment of the health of Non-woodland trees

10. Life expectancy tables

11. Rates of diameter growth of Non-woodland trees

APPENDIX 1

Glossary of terms and abbreviations

Approved woodland

Privately owned woodland included in a Forestry Commission scheme where the owners could not, or did not wish to, enter into the long-term, legally-binding arrangement of Dedication.

Broadleaved high forest of coppice origin

Crops of coppice origin which have a mean breast height diameter of more than 15 cm and are assessed by the same criteria as Broadleaved High Forest.

Cleared

Woodland areas which are marked green on the O.S. 1:50000 map, but at the time of survey were found to be cleared of trees and had not been converted to another land use.

Clump

A small woodland or group of trees of less than 0.25 ha.

Coppice

Crops of marketable broadleaved species that have at least two stems per stool and are either being worked or are capable of being worked on rotation. With the exception of hazel coppice, more than half the stems should be capable of producing 3 m timber lengths of good form. Coppice crops with a mean breast height diameter greater than 15 cm are assessed as Broadleaved High Forest of Coppice origin.

Coppice with standards

Two-storey stands where the overstorey consists of at least 25 stems per hectare that are older than the understorey of worked Coppice by at least one Coppice rotation.

Dedicated woodland

Privately owned woodland within the Forestry Commission Dedication Scheme. In return for financial assistance, an owner accepts a continuing obligation by Deed or Agreement of Covenant to manage these woodlands in accordance with a Plan of Operations which is designed to secure sound forestry practice.

Diameter at breast height (dbh)

Diameter of a tree rounded down to the nearest centimetre at a point on the tree 1.3 m above ground level.

Disforested

Woodland areas which are marked green on the O.S. 1:50000 map, but at the time of survey were found to be under another land use, e.g. agriculture, buildings.

Extra woodland

Areas of woodland over 0.25 of a hectare in extent found during the Census but which were not marked green on the O.S. 1:50000 map.

Forestry Commission woodland

Woodland owned by, on lease to or managed by the Forestry Commission.

High forest

Stands of trees having a canopy density of 20 per cent or more, or, in the case of young stands which have not closed canopy, occupying 20 per cent or more of the ground at normal spacing. More than half of the crop should be capable of producing 3 m timber lengths of good form and be of merchantable species.

Linear feature

A strip of woody vegetation less than 20 m mean width, crown edge to crown edge and more than 25 m long.

Mainly broadleaved high forest

High Forest (q.v.) containing 50 per cent or more by area of broadleaved species.

Mainly coniferous high forest

High forest (q.v.) containing more than 50 per cent by area of coniferous species.

'Other' woodland

Woodland which is neither in Forestry Commission ownership or management nor included in a Dedication or Approved Woodland Scheme.

Planting year (P year)

The year in which the trees were planted or regenerated naturally. With older crops it was often necessary to estimate the P Year.

Planting year class

A group of planting years.

Scrub

All inferior crops where more than half the trees are of poor form, poor timber potential or composed of unmarketable species, and so do not qualify as either High Forest or Coppice.

Shrub layer

A layer of woody plants below the tree canopy.

Woodland

Area of woody growth greater than 0.25 of a hectare in area and at least 20 m wide. Where the stocking density was less than 20 per cent or there was evidence of recent woody growth, the area was described as Cleared, otherwise it was allocated to a forest type.

APPENDIX 2

List of species recorded and reported upon – Woodland

English name	Standard abbreviation	Botanical name
Scots pine	SP	*Pinus sylvestris* L.
Corsican pine	CP	*Pinus nigra* var. *maritima* (Ait.) Melville
Lodgepole pine	LP	*Pinus contorta* Douglas ex Loud.
Sitka spruce	SS	*Picea sitchenis* (Bong.) Carr.
Norway spruce	NS	*Picea abies* (L.) Karst.
European larch	EL	*Larix decidua* Miller
Japanese/Hybrid larch	JL	*Larix kaempferi* (Lamb.) Carr.
	HL	*Larix x eurolepis* Henry
Douglas fir	DF	*Pseudotsuga menziesii* (Mirb.) Franco
Other conifers*	XC	
Mixed conifers	MC	
Oak	OK	*Quercus robur* L.
		Quercus petraea (Matt.) Lieblein.
Beech	BE	*Fagus sylvatica* L.
Sycamore	SY	*Acer pseudoplatanus* L.
Ash	AH	*Fraxinus excelsior* L.
Birch	BI	*Betulus* spp.
Poplar	PO	*Populus* spp.
Sweet chestnut	SC	*Castanea sativa* Mill.
Alder	AR	*Alnus* spp.
Elm	EM	*Ulmus* spp.
Hornbeam	HBM	*Carpinus betulus* L.
Hazel	HAZ	*Corylus avellana* L.
Willow		*Salix* spp.
Other broadleaves*	XB	
Mixed broadleaves	MB	

* Included within Other conifers and Other broadleaves are some species which were recognised in the Census but are of such limited occurrence as to preclude their individual inclusion in the Report.

List of species recorded but not reported upon individually – Woodland

English name	Standard abbreviation	Botanical name
Other pine	XP	*Pinus* spp.
Other spruce	XS	*Picea* spp.
Western hemlock	WH	*Tsuga heterophylla* (Raf.) Sarg.
Western red cedar	RC	*Thuja plicata* D. Don
Cypresses		*Cupressus* spp.
		Chamaecyparis spp.
		x *Cupressocyparis leylandii* (Jacks. Dallim.) Dallim.
Grand fir	GF	*Abies grandis* Lindl.
Noble fir	NF	*Abies procera* Rehd.
Other fir	XF	*Abies* spp.
Redwoods		*Sequoia sempervirens* (D. Don.) Endl.
		Sequoiadendron giganteum (Lindl.) Buchholz
Yew		*Taxus baccata* L.
Other conifers	XC	
Other oak		*Quercus* spp.
Norway maple	NOM	*Acer platanoides* L.
Horse chestnut	HCH	*Aesculus hippocastanum* L.
Lime	LI	*Tilia* spp.
English elm**	EEM	*Ulmus procera* Salis.
Wych elm**	WEM	*Ulmus glabra* Huds.
Nothofagus	N	*Nothofagus* spp.
Prunus (Cherries)		*Prunus* spp.
Ornamentals		
Other broadleaves	XB	

**For the purpose of the Report, English elm and Wych elm were included as elm.

Note:
In certain circumstances the following were also recorded as Woodland species:
Rowan *Sorbus aucuparia* L.
Holly *Ilex aquifolium* L.
Field maple *Acer campestre* L.
Whitebeam *Sorbus aria* agg.

H

List of species recorded and reported upon – Non-woodland trees

English name	Standard abbreviation	Botanical name
Pines		*Pinus* spp.
Spruces		*Picea* spp.
Larches		*Larix* spp.
Cypresses		*Cupressus* spp.
		Chamaecyparis spp.
		x *Cupressocyparis leylandii* (Jacks. Dallim.) Dallim.
Other conifers	XC	
Oak	OK	*Quercus robur* L.
		Quercus petraea (Matt.) Lieblein.
Beech	BE	*Fagus sylvatica* L.
Sycamore	SY	*Acer pseudoplatanus* L.
Ash	AH	*Fraxinus excelsior* L.
Birch	BI	*Betula* spp.
Poplar	PO	*Populus* spp.
Sweet chestnut	SC	*Castanea sativa* Mill.
Horse chestnut	HCH	*Aesculus hippocastanum* L.
Alder	AR	*Alnus* spp.
Lime	LI	*Tilia* spp.
Elm	EM	*Ulmus* spp.
Willow		*Salix* spp.
Other broadleaves	XB	

Note:
Although the above species are given in the Report the total list of species recorded was the same as for Woodland.

List of shrub layer species recorded

English name	Botanical name
Rowan*	*Sorbus aucuparia* L.
Field maple*	*Acer campestre* L.
Blackthorn	*Prunus spinosa* L.
Hawthorn	*Crataegus monogyna* Jacq.
Rhododendron	*Rhododendron* spp.
Holly*	*Ilex aquifolium* L.
Elder	*Sambucus* spp.
Broom	*Sarothamnus scoparius* (L.) Wimmer ex Koch.
Gorse	*Ulex* spp.
Privet	*Ligustrum vulgare* L.
Dogwood	*Cornus sanguinea* L.
Sallow	*Salix caprea* L.
Box	*Buxus sempervirens* L.
Whitebeam*	*Sorbus aria* agg.
Spindle	*Euonymus europaeus* L.
Yew*	*Taxus baccata* L.
Hornbeam*	*Carpinus betulus* L.
Hazel	*Corylus avellana* L.
Willow*	*Salix* spp.
Other shrubs	
Mixed shrubs	

* These species have on occasion been recognised as tree species.

APPENDIX 3

Woodland type distribution by county, region and country

Hectares

County or region	Mainly coniferous high forest	Mainly broadleaved high forest	Total high forest	Coppice with standards	Coppice	Scrub	Cleared	Total
England								
Avon	1 509	4 609	6 118	—	142	442	160	6 862
Bedfordshire	1 748	3 613	5 361	4	14	544	261	6 184
Berkshire	5 754	9 447	15 201	143	321	954	399	17 018
Buckinghamshire	3 511	10 790	14 301	6	9	1 010	292	15 618
Cambridgeshire	1 072	4 814	5 886	26	—	382	318	6 612
Cheshire	2 310	6 329	8 639	13	—	112	78	8 842
Cleveland	1 678	1 493	3 171	—	—	162	28	3 361
Cornwall	5 110	6 892	12 002		30	488	5 252	529
Cumbria	35 340	13 806	49 146	89	134	4 196	885	54 450
Derbyshire	4 322	7 239	11 561	—	3	1 656	79	13 299
Devon	21 399	22 804	44 203	22	89	8 504	995	53 813
Dorset	10 660	10 134	20 794	624	241	2 639	764	25 062
Durham	7 801	5 358	13 159	—	—	599	176	13 934
East Sussex	7 075	10 976	18 051	2 121	4 438	3 063	371	28 044
Essex	2 097	10 934	13 031	721	681	307	256	14 996
Gloucestershire	11 575	12 530	24 105	71	322	1 301	573	26 372
Greater London	202	5 100	5 302	—	—	686	44	6 032
Greater Manchester	214	2 092	2 306	—	1	294	48	2 649
Hampshire	19 230	34 940	54 170	76	601	6 575	627	62 049
Hereford & Worcester	10 272	16 468	26 740	66	636	714	601	28 757
Hertfordshire	2 216	7 334	9 550	189	357	1 488	786	12 370
Humberside	2 880	5 822	8 702	—	—	680	235	9 617
Isle of Wight	1 083	2 027	3 110	10	17	519	39	3 695
Kent	6 327	14 056	20 383	4 340	13 574	3 127	1 240	42 664
Lancashire	3 852	6 587	10 439	10	29	835	164	11 477
Leicestershire	879	5 536	6 415	7	8	1 258	53	7 741
Lincolnshire	6 109	11 356	17 465	30	36	925	318	18 774
Merseyside	313	1 343	1 656	—	—	14	7	1 677
Norfolk	21 350	15 397	36 747	363	60	4 579	948	42 697
North Yorkshire	33 176	19 700	52 876	—	6	2 510	460	55 852
Northamptonshire	3 997	6 515	10 512	263	122	455	557	11 909
Northumberland	66 231	6 569	72 800	—	5	1 871	1 094	75 770
Nottinghamshire	7 389	6 535	13 924	16	15	555	548	15 058
Oxfordshire	3 644	10 478	14 122	149	54	637	407	15 369
Shropshire	11 250	12 367	23 617	16	60	1 052	530	25 275
Somerset	7 971	7 406	15 377	9	34	3 301	579	19 300
South Yorkshire	2 305	5 439	7 744	—	—	2 438	216	10 398
Staffordshire	6 783	8 334	15 117	6	448	1 027	338	16 936
Suffolk	13 406	13 507	26 913	101	85	588	524	28 211
Surrey	7 410	16 260	23 670	706	1 039	5 362	754	31 531
Tyne & Wear	776	810	1 586	—	—	98	6	1 690
Warwickshire	1 489	4 922	6 411	4	7	284	177	6 883
West Midlands	286	1 332	1 618	—	—	114	95	1 827

APPENDIX 3 – cont.

Hectares

County or region	Mainly coniferous high forest	Mainly broadleaved high forest	Total high forest	Coppice with standards	Coppice	Scrub	Cleared	Total
West Sussex	8 448	18 466	26 914	871	1 426	4 468	845	34 524
West Yorkshire	1 689	6 828	8 517	—	—	651	182	9 350
Wiltshire	8 361	13 949	22 310	372	209	1 273	675	24 839
Total England	382 497	429 248	811 745	11 473	25 711	79 498	19 261	947 688
Wales								
Clwyd	14 584	6 391	20 975	—	114	658	247	21 994
Dyfed	40 694	14 709	55 403	36	1 422	2 678	587	60 126
Gwent	8 888	6 489	15 377	27	39	1 318	139	16 900
Gwynedd	31 586	9 048	40 634	—	20	1 602	334	42 590
Mid Glamorgan	11 345	3 143	14 488	—	103	343	340	15 274
Powys	47 031	15 299	62 330	6	28	1 092	1 248	64 704
South Glamorgan	708	1 512	2 220	11	39	127	67	2 464
West Glamorgan	13 124	2 730	15 854	—	85	404	390	16 733
Total Wales	167 960	59 321	227 281	80	1 849	8 222	3 352	240 784
Scotland								
Borders	58 875	7 055	65 930	4	3	1 015	979	67 931
Central	29 137	3 810	32 947	—	—	2 691	656	36 294
Dumfries & Galloway	124 935	8 112	133 047	4	1	1 178	1 390	135 620
Fife	8 975	2 751	11 726	—	—	1 102	685	13 513
Grampian	109 680	11 959	121 639	1	—	6 102	3 170	130 912
Highland	195 078	10 475	205 553	—	—	29 733	4 398	239 684
Lothian	7 921	5 071	12 992	6	—	675	448	14 121
Strathclyde	175 299	17 262	192 561	—	—	13 361	2 845	208 767
Tayside	56 451	9 354	65 805	—	—	4 655	2 623	73 083
Total Scotland	766 351	75 849	842 200	15	4	60 512	17 194	919 925

Note: The figures may not add to totals due to rounding errors.

APPENDIX 4
Woodland density by county, region and country

County	Woodland area (hectares)	Land & inland water area (hectares)‡	Woodland density (per cent)
England			
Avon	6 862	134 605	5.1
Bedfordshire	6 184	123 467	5.0
Berkshire	17 018	125 891	13.5
Buckinghamshire	15 618	188 285	8.3
Cambridgeshire	6 612	340 901	1.9
Cheshire	8 842	232 846	3.8
Cleveland	3 361	58 307	5.8
Cornwall	18 301	356 422	5.1
Cumbria	54 450	681 015	8.0
Derbyshire	13 299	263 096	5.1
Devon	53 813	671 087	8.0
Dorset	25 062	265 380	9.4
Durham	13 934	243 591	5.7
East Sussex	28 044	179 519	15.6
Essex	14 996	367 188	4.1
Gloucestershire	26 372	264 263	10.0
Greater London	6 032	157 946	3.8
Greater Manchester	2 649	128 674	2.1
Hampshire	62 049	377 685	16.4
Hereford and Worcester	28 757	392 648	7.3
Hertfordshire	12 370	163 418	7.6
Humberside	9 617	351 226	2.7
Isle of Wight	3 695	38 097	9.7
Kent	42 664	373 063	11.4
Lancashire	11 477	306 347	3.7
Leicestershire	7 741	255 293	3.0
Lincolnshire	18 774	591 484	3.2
Merseyside	1 677	65 202	2.6
Norfolk	42 697	536 823	7.9
North Yorkshire	55 852	830 869	6.7
Northamptonshire	11 909	236 734	5.0
Northumberland	75 770	503 166	15.1
Nottinghamshire	15 058	216 365	7.0
Oxfordshire	15 369	260 793	6.0
Shropshire	25 275	349 014	7.2
Somerset	19 300	345 043	5.6
South Yorkshire	10 398	156 049	6.7
Staffordshire	16 936	271 616	6.2
Suffolk	28 211	379 667	7.4
Surrey	31 531	167 924	18.8
Tyne and Wear	1 690	54 005	3.1
Warwickshire	6 883	198 053	3.5
West Midlands	1 827	89 942	2.0
West Sussex	34 524	198 939	17.4
West Yorkshire	9 350	203 912	4.6
Wiltshire	24 839	348 070	7.1
Total England	947 688	13 043 927	7.3

‡ As at 31 May 1978. (*Source*: Ordnance Survey.)

111

APPENDIX 4 – *cont.*

Woodland density by county, region and country

County	Woodland area (hectares)	Land & inland water area (hectares)‡	Woodland density (per cent)
Wales			
Clwyd	21 994	242 645	9.1
Dyfed	60 126	576 577	10.4
Gwent	16 900	137 600	12.3
Gwynedd	42 590	386 686	11.0
Mid Glamorgan	15 274	101 867	15.0
Powys	64 704	507 740	12.7
South Glamorgan	2 464	41 630	5.9
West Glamorgan	16 733	81 657	20.5
Total Wales	240 784	2 076 402	11.6
Scotland (Region)			
Borders	67 931	469 836	14.5
Central	36 294	270 489	13.4
Dumfries & Galloway	135 620	642 545	21.1
Fife	13 513	131 734	10.3
Grampian	130 912	875 303	15.0
Highland	239 684	2 613 639	9.2
Lothian	14 121	176 740	8.0
Strathclyde	208 767	1 376 789	15.2
Tayside	73 083	764 189	9.6
*Total Scotland	919 925	7 321 263	12.6

* Omitting Western Isles, Orkneys and Shetlands.
‡ As at 31 May 1978 for Wales and at 31 December 1977 for Scotland. (*Source*: Ordnance Survey.)

APPENDIX 5
Predominant coniferous and broadleaved species in high forest by county, region and country

County	Predominant conifer	Predominant broadleaved
England		
Avon	Jap./Hybrid larch	Ash
Bedfordshire	Scots pine	Oak
Berkshire	Scots pine	Oak
Buckinghamshire	Norway spruce	Beech
Cambridgeshire	Norway spruce	Ash
Cheshire	Scots pine	Oak
Cleveland	Scots pine	Sycamore
Cornwall	Sitka spruce	Oak
Cumbria	Sitka spruce	Oak
Derbyshire	Scots pine	Oak
Devon	Sitka spruce	Oak
Dorset	Corsican pine	Oak
Durham	Sitka spruce	Sycamore
East Sussex	Scots pine	Oak
Essex	Scots pine	Oak
Gloucestershire	Norway spruce	Oak
Greater London	Scots pine	Oak
Greater Manchester	Corsican pine	Oak
Hampshire	Scots pine	Oak
Hereford & Worcester	Douglas fir	Oak
Hertfordshire	Scots pine	Oak
Humberside	Scots pine	Sycamore
Isle of Wight	Corsican pine	Oak
Kent	Scots pine	Oak
Lancashire	Sitka spruce	Sycamore
Leicestershire	Norway spruce	Oak
Lincolnshire	Scots pine	Ash
Merseyside	Scots pine	Sycamore
Norfolk	Scots pine	Oak
North Yorkshire	Scots pine	Oak
Northamptonshire	Norway spruce	Oak
Northumberland	Sitka spruce	Birch
Nottinghamshire	Corsican pine	Birch
Oxfordshire	European larch	Beech
Shropshire	Douglas fir	Oak
Somerset	Sitka spruce	Oak
South Yorkshire	Corsican pine	Sycamore
Staffordshire	Scots pine	Birch
Suffolk	Scots pine	Oak
Surrey	Scots pine	Oak
Tyne and Wear	Corsican pine	Sycamore
Warwickshire	Scots pine	Oak
West Midlands	Jap./Hybrid larch	Oak
West Sussex	Scots pine	Oak
West Yorkshire	Jap./Hybrid larch	Oak
Wiltshire	Jap./Hybrid larch	Oak
England	Scots pine	Oak

APPENDIX 5 – *cont.*

County	Predominant conifer	Predominant broadleaved
Wales		
Clwyd	Sitka spruce	Oak
Dyfed	Sitka spruce	Oak
Gwent	Jap./Hybrid larch	Oak
Gwynedd	Sitka spruce	Oak
Mid Glamorgan	Sitka spruce	Oak
Powys	Sitka spruce	Oak
South Glamorgan	Jap./Hybrid larch	Oak
West Glamorgan	Sitka spruce	Oak
Wales	Sitka spruce	Oak
Scotland (Region)		
Borders	Sitka spruce	Oak
Central	Sitka spruce	Oak
Dumfries & Galloway	Sitka spruce	Oak
Fife	Scots pine	Birch
Grampian	Scots pine	Birch
Highland	Scots pine	Birch
Lothian	Sitka spruce	Oak
Strathclyde	Sitka spruce	Oak
Tayside	Sitka spruce	Birch
Scotland	Sitka spruce	Birch

APPENDIX 6

Description of soil strata

Reassessment of Soil Survey of England and Wales 1:1 million soil map for woodland surveys

The Forestry Commission Census Section sought advice from the Soil Survey of England and Wales and the Forestry Commission's own Site Studies Branch on soils to be recognised for Census purposes. The object was to produce a map showing broad site types relevant to tree growth potential. As a result, the 71 soil units shown on the 1:1 million soil map were combined to produce 16 soil groups.

A further variable was recognised which overrode the new soil strata units: potential soil moisture deficit (PSMD). Deficits more or less than 150mm were distinguished at county level; counties with >150mm PSMD were considered dry (namely, those east of and including Nottinghamshire, Lincolnshire, Leicestershire, Northamptonshire, Oxfordshire, Berkshire and Hampshire), and counties in the 100–150mm zone were intermediate. Wet uplands (<100mm PSMD) were already separated (units 12 to 16 in the list).

Soil groups

1. Sandy; well drained.
2. Alluvial and 'valley' soils; with groundwater.
3. Lowland peaty and humose soils; with groundwater.
4. Rendzinas over chalk and limestone; well drained.
5. Brown calcareous soils; well drained.
6. Lowland brown earths; mainly well drained.
7. Deeply leached brown earths; mainly over chalk.
8. Podzols; well drained.
9. Sandy soils, some podzolisation, with groundwater.
10. Surface-water gleys and other clayey soils.
11. Surface-water gleys over compacted silty or loamy beds. (High Weald.)
12. Brown earths; uplands.
13. Stagno-podzols; humose or peaty; often with iron-pan and rock.
14. Surface-water gleys in moist climates.
15. Peaty or humose surface-water gleys.
16. Hill peat.
17. Urban.

Reassessment of the Macaulay Institute for Soil Research 1:625 000 soil map for woodland surveys

Discussions with staff of the Macaulay Institute for Soil Research made it apparent that, although many of the soil groups adopted for England and Wales could also be utilized for Scotland, there was a need for some additional groups. After further advice from Site Studies Branch, seven additional groupings were adopted. Some of these are best described as complexes to cater for changes in soil type that take place over short distances.

18. Western seaboard complex.
19. Peaty gley complex.
20. West coast igneous peaty complex.
21. North and west coast rock and peat complex.
22. Blanket peatlands.
23. Mountain tops.
24. Coarse textured alluvium; high groundwater.

APPENDIX 7
Distribution of soil groups by county in England and Wales and by Conservancy in Scotland

County								Soil groups									
	1	2	3	4	5	6	7	8	9	10	11	12	13	14	15	16	17
England																	
Avon		•		•		•				•		•					
Bedfordshire	•	•		•	•	•	•			•							
Berkshire		•		•		•			•	•							
Buckinghamshire		•		•	•	•	•			•							
Cambridgeshire	•	•	•	•						•							
Cheshire	•	•	•					•	•	•		•	•	•	•		
Cleveland						•				•				•			
Cornwall	•					•			•			•	•	•		•	
Cumbria	•	•	•		•	•		•	•	•		•	•	•	•	•	
Derbyshire	•	•	•		•	•		•	•	•		•	•	•	•	•	
Devon	•	•				•	•	•	•	•		•	•	•	•	•	
Dorset	•	•	•	•		•	•	•		•	•	•					
Durham	•					•				•				•	•	•	
East Sussex	•	•		•						•	•						
Essex		•	•		•	•				•							
Gloucestershire		•		•	•	•	•			•		•			•		
Greater London				•			•	•		•							•
Greater Manchester																	•
Hampshire	•	•		•		•	•	•	•	•							
Hereford & Worcester	•	•								•		•			•		
Hertfordshire		•		•	•	•	•			•							
Humberside	•	•	•	•		•			•	•							
Isle of Wight				•		•				•							
Kent	•	•		•		•	•			•	•						
Lancashire	•	•	•		•	•			•	•		•	•	•	•	•	
Leicestershire		•		•	•	•				•							
Lincolnshire		•	•	•		•			•	•							
Merseyside																	•
Norfolk	•	•		•		•				•							
North Yorkshire	•			•	•	•				•		•	•	•	•	•	
Northamptonshire		•		•	•	•				•							
Northumberland	•					•				•			•	•	•	•	
Nottinghamshire	•	•	•					•	•	•		•		•			
Oxfordshire		•		•	•		•			•		•			•		
Shropshire	•	•	•			•		•	•	•		•	•	•	•		
Somerset	•	•	•			•	•			•		•	•	•	•	•	
South Yorkshire	•	•	•		•	•			•	•		•	•	•	•	•	•
Staffordshire	•	•	•			•			•	•		•	•	•	•		
Suffolk	•		•		•	•				•							
Surrey	•	•		•		•	•		•	•	•						•
Tyne and Wear						•				•							
Warwickshire		•				•				•		•			•		
West Midlands																	•
West Sussex	•	•		•		•	•	•	•	•	•						
West Yorkshire			•		•	•				•		•	•	•	•	•	•
Wiltshire	•	•		•		•	•	•		•							

116

APPENDIX 7 – *contd.*

County							Soil groups										
	1	2	3	4	5	6	7	8	9	10	11	12	13	14	15	16	17
Wales																	
Clwyd	●	●	●			●				●	●		●			●	
Dyfed	●	●				●						●	●	●	●	●	
Gwent		●				●				●		●			●		
Gwynedd	●					●						●		●	●	●	
Mid Glamorgan	●					●						●			●		
Powys		●				●				●		●	●	●	●	●	●
South Glamorgan	●	●				●						●					
West Glamorgan						●						●		●	●		

Conservancy									Soil groups															
	1	2	3	4	5	6	7	8	9	10	11	12	13	14	15	16	17	18	19	20	21	22	23	24
Scotland																								
North Scotland	●												●	●				●	●	●	●	●	●	●
East Scotland	●	●				●						●	●	●		●			●			●	●	●
South Scotland	●	●				●						●	●	●	●	●	●		●		●		●	●
West Scotland	●	●				●						●	●	●	●		●	●	●	●	●	●	●	●

APPENDIX 8

Non-woodland tree information by county, region and country

County or Region	Total number of all live trees (thousands)	Total number of clumps	Total length of linear features (km)	Density per square kilometre		
				Number of live trees	Number of clumps	Length of linear features (km)
England						
Avon	743	78 700	760	552	58	0.6
Bedfordshire	896	19 550	580	726	16	0.5
Berkshire	804	51 280	960	639	41	0.8
Buckinghamshire	891	61 610	1 020	473	33	0.5
Cambridgeshire	1 188	27 340	690	348	8	0.2
Cheshire	2 030	106 270	3 440	872	46	1.5
Cleveland	120	4 000	110	206	7	0.2
Cornwall	1 576	142 490	1 060	442	40	0.3
Cumbria	4 419	155 360	2 710	649	23	0.4
Derbyshire	1 130	75 330	390	429	29	0.1
Devon	7 733	413 010	5 260	1 152	62	0.8
Dorset	804	57 170	1 270	303	22	0.5
Durham	319	19 890	330	131	8	0.2
East Sussex	1 105	35 030	2 850	615	20	1.6
Essex	2 379	174 620	1 540	648	48	0.4
Gloucestershire	1 355	132 680	1 910	513	50	0.7
Greater London	1 648	69 460	1 270	1 043	44	0.8
Greater Manchester	1 877	109 910	1 130	1 459	85	0.9
Hampshire	3 877	246 240	2 850	1 027	65	0.8
Hereford & Worcester	2 657	177 530	2 350	677	45	0.6
Hertfordshire	1 107	41 250	810	677	25	0.5
Humberside	699	44 060	310	199	13	0.1
Isle of Wight	161	5 600	300	423	15	0.8
Kent	1 666	77 690	5 480	446	21	1.4
Lancashire	1 532	92 490	2 070	500	30	0.7
Leicestershire	1 258	80 480	1 220	493	32	0.5
Lincolnshire	1 521	59 400	680	257	10	0.1
Merseyside	720	31 830	530	1 105	49	0.8
Norfolk	1 489	77 020	1 170	277	14	0.2
North Yorkshire	3 501	189 280	4 250	421	23	0.5
Northamptonshire	670	31 350	530	283	13	0.2
Northumberland	812	33 810	990	161	7	0.2
Nottinghamshire	495	14 530	150	229	7	0.1
Oxfordshire	1 524	76 280	1 930	584	29	0.7
Shropshire	2 192	85 730	1 960	628	25	0.6
Somerset	2 671	188 480	1 550	774	55	0.4
South Yorkshire	2 041	161 080	690	1 308	103	0.4
Staffordshire	2 116	73 340	2 970	779	27	1.1
Suffolk	2 666	132 430	2 290	702	35	0.6
Surrey	2 230	96 930	2 940	1 328	58	1.8
Tyne and Wear	177	3 880	100	328	7	0.2
Warwickshire	1 495	74 230	1 150	755	37	0.6
West Midlands	1 266	75 250	450	1 407	84	0.5

118

APPENDIX 8 – *cont.*

County or Region	Total number of all live trees (thousands)	Total number of clumps	Total length of linear features (km)	Density per square kilometre		
				Number of live trees	Number of clumps	Length of linear features (km)
West Sussex	961	37 400	1 260	483	19	0.6
West Yorkshire	1 206	71 770	1 040	591	35	0.5
Wiltshire	1 689	176 630	1 000	485	51	0.3
Total England	75 415	4 189 690	70 300	578	32	0.5
Wales						
Clwyd	2 439	113 730	1 900	1 005	47	0.8
Dyfed	4 330	96 930	7 650	751	17	1.3
Gwent	1 266	66 220	1 320	920	48	1.0
Gwynedd	2 742	157 710	1 800	709	41	0.5
Mid Glamorgan	709	17 880	700	696	18	0.7
Powys	3 202	190 950	2 840	631	38	0.6
South Glamorgan	262	8 050	240	629	19	0.6
West Glamorgan	576	14 840	560	705	18	0.7
Total Wales	15 526	666 310	17 010	748	32	0.8
Scotland						
Borders	1 340	47 410	820	285	10	0.2
Central	1 046	67 190	760	387	25	0.3
Dumfries & Galloway	1 178	55 290	1 000	183	9	0.2
Fife	637	18 680	620	484	14	0.5
Grampian	3 674	148 680	3 350	420	17	0.4
Highland	1 882	134 490	980	72	5	<0.1
Lothian	1 188	26 400	490	672	15	0.3
Strathclyde	4 671	250 100	3 110	339	18	0.2
Tayside	1 896	71 900	1 760	248	9	0.2
Total Scotland	17 513	820 140	12 890	239	11	0.2
Total Great Britain	108 454	5 676 140	100 200	483	25	0.4

119

APPENDIX 9

Criteria for the assessment of the health of non-woodland trees

For all living trees, health was estimated in three categories; good, moderate and poor. Symptoms of poor health were:

a. Crown deterioration, indicated by:

abnormally small, sparse or unhealthily discoloured foliage;

premature discolouration of foliage or defoliation;

extensive dieback, breakage or shedding of limbs in the upper crown (disregarding 10 per cent of dieback in oak).

b. Bole deterioration, indicated by:

diseased, dead or missing areas of bark including decayed wood;

death of large limbs;

advanced and hazardous decay following lopping;

suspected internal decay of swollen boles.

c. Instability, indicated by:

wind-rock symptoms of displaced soil at the base of the bole;

exposure of root system through erosion.

From an assessment of the presence or otherwise of any of the above symptoms the condition of each tree was classified as good, moderate or poor. All assessments were external from ground level.

If none of the above symptoms were present, the health of the tree was assessed as 'good'.

If one symptom only was present, the health was assessed as 'moderate'.

If more than one symptom was present, tree health was assessed as 'poor'.

There were occasions, particularly in summer, when the general appearance of a tree was unsatisfactory, and then the surveyor, if in doubt, recorded tree health as 'moderate'.

Notes:

1. Dead branches or areas of dead bark in beech automatically classified the tree health as 'poor'.
2. Elm was treated on its own, as the symptoms of Dutch elm disease can occur very quickly during the latter part of the summer. For prognosis, the general health of the tree was compared with those around it. Checks were made for dead leaves, twigs, branches and 'shepherds crooks' as well as for beetle emergence holes in the bark. Areas of dead or peeling bark indicated serious loss of health.

APPENDIX 10

a) Life expectancy by species groups, size and health of coniferous species reported upon in the non-woodland tree tables

Years

Species groups	Pines Larches			Spruces Douglas fir Other firs			Other conifers		
Health	Good	Mod.	Poor	Good	Mod.	Poor	Good	Mod.	Poor
Size class (dbh)									
7–20 cm	180	90	40	150	70	30	100	60	30
21–50 cm	100	60	20	90	50	20	80	50	20
51–80 cm	80	50	10	70	40	10	60	40	10
>80 cm	50	20	—	40	20	—	40	20	—

Notes:

The object of this table is to give a broad assessment of life expectancy of non-woodland trees, thus allowing forecasts to be made of the likely changes in the tree population in the landscape.

The years of life expectancy are broad national figures, assuming normal conditions over the period. Allowance must be made for local climatic and soil conditions.

Redwoods and yew in good health must be considered separately from all other conifers as both species may live a very considerable time.

b) Life expectancy by species groups, size and health of broadleaved species reported upon in the non-woodland tree tables

Years

Species groups	Oak Sweet chestnut			Sycamore Lime			Beech* Elm*		
Health	Good	Mod.	Poor	Good	Mod.	Poor	Good	Mod.	Poor
Size class (dbh)									
7–20 cm	>300	>250	100	>200	>150	60	>150	80	—
21–50 cm	>250	>200	90	>200	>100	50	>100	60	—
51–80 cm	>200	>150	70	>150	>100	40	90	40	—
>80 cm	>150	>100	50	>100	>80	20	60	30	—

Species groups	Ash			Horse chestnut Willow, Alder, Poplar			Birch		
Health	Good	Mod.	Poor	Good	Mod.	Poor	Good	Mod.	Poor
Size class (dbh)									
7–20 cm	100	70	30	80	50	20	50	20	—
21–50 cm	80	50	20	50	30	10	30	10	—
51–80 cm	60	40	10	30	10	—	20	—	–-
>80 cm	40	30	—	10	—	—	10	—	—

Notes:

The object of this table is to give a broad assessment of life expectancy of non-woodland trees, thus allowing forecasts to be made of the likely changes in the tree population in the landscape.

The years of life expectancy are broad national figures, assuming normal conditions over the period. Allowance must be made for local climatic and soil conditions.

The species in this table are only those found in Table 18c of the individual county, Conservancy and country reports. The species in the 'Other broadleaves' category, namely those recognised in the Non-Woodland Tree Survey, but not reported upon individually in Table 18c, cover such a wide range of life expectancies that it is not feasible to classify them in any meaningful way.

* In the case of elms, consideration must be given to the local incidence of Dutch elm disease. The life expectancy stated above assumes that the trees will be free of this particular threat. No life expectancy is given for beech or elm in poor health because of the likelihood of Beech bark disease and Dutch elm disease; such trees may live for some time or be dead next year.

APPENDIX 11

Rates of diameter growth of non-woodland trees

With broadleaved species featuring so widely in the countryside in the form of Isolated Trees, Clumps and Linear Features concern has been expressed that the distribution of tree numbers throughout the diameter classes may be inadequate in some areas to maintain the present growing stock, and in others could lead to future imbalance. In order to be able to predict future changes in the number of trees in the non-woodland category it is first necessary to know the average time it will take for Isolated Trees and trees in Clumps and Linear Features to grow through the diameter classes.

Competition for growing space tends to encourage height growth in a tree at the expense of diameter growth. Consequently Isolated Trees, which by definition are not subject to competition from neighbouring trees, and are of a given age can be expected to increase their breast height diameter at a somewhat faster rate than trees of the same age in Clumps and Linear Features where some degree of neighbouring competition exists. They, in turn, can be expected to grow somewhat faster than trees of similar age in woodland where competition is continuous throughout the rotation.

Such information as we have makes it reasonable to assume that the rate of diameter increase in trees in Clumps and Linear Features can be equated with that of the 100 largest trees per hectare in woodland, i.e. the fast growing stems which have already assumed dominance and can be expected to maintain that dominance throughout their lives. It can further be assumed that the diameter growth of Isolated Trees will be even faster.

In order to cater for the wide range of rates of growth which occur throughout Great Britain the following table shows the number of years the mean tree of a range of species and yield classes will take to grow through the diameter classes under woodland conditions and also under the more limited competition of Clumps and Linear Features. Much less is known about the rates of diameter increase of free grown trees and about the range of values that can be expected between good and poor sites; consequently, only one set of average values has been given for this category.

From this table it will be seen that for oak of average growth it will take about 20 years for the mean tree in a woodland crop to attain 7 cm dia-

Table 53 Number of years it will take trees of selected species to grow through each diameter class

Dbh class cm	Oak			Beech				Sycamore, Ash and Birch					Poplar 7.3 m × 7.3 m					
	YC 8	YC 6	YC 4	YC 10	YC 8	YC 6	YC 4	YC 12	YC 10	YC 8	YC 6	YC 4	YC 14	YC 12	YC 10	YC 8	YC 6	YC 4
Woodland trees																		
>7	15	20	25	18	21	27	34	9	11	13	15	18	3	3	4	4	5	6
7–20	25	29	35	24	29	32	36	13	14	16	22	32	6	7	7	8	9	11
21–30	20	23	27	15	17	19	22	7	9	11	21	50	6	6	7	8	9	12
31–50	35	47	90	32	38	45	58	21	31	60	—	—	15	17	21	32	56	—
Hedgerows and clumps																		
>7	12	13	16	12	12	16	19	6	7	8	10	12	3	3	4	4	5	6
7–20	25	29	34	21	25	28	34	12	13	15	19	25	6	7	7	8	9	11
21–30	18	22	30	12	15	17	21	7	9	11	19	39	5	5	6	7	8	10
31–50	30	45	85	31	37	44	55	21	30	56	—	—	11	13	17	24	33	50
Open grown trees																		
>7		10			12					6						4		
7–20		17			19					13						8		
21–30		15			14					10						7		
31–50		35			30					30						20		

meter at breast height. In a Clump or Linear Feature one could expect this time to be reduced to about 13 years, and for an Isolated Tree to about 10 years. Consequently, whereas it will take nearly 120 years for the mean tree in a normally thinned woodland crop to attain 50 cm dbh it will take only about 110 years if it is in a Clump and only about 80 years if it occurs as an Isolated Tree.

It must be stressed that the Table shows average times of passage through the diameter classes and there will obviously be considerable variation around these values depending on local factors. On poorer sites, for example, while oak can almost always attain 50 cm dbh eventually, it is doubtful if sycamore, ash or birch can do so as their life span on these sites is likely to be too short. It must also be pointed out that a further estimate, namely that of losses, both natural and from clear felling, must be built-in to the tree numbers in each class before it is possible to calculate whether the current tree numbers are adequate to maintain, or even improve, the pattern of distribution for the future. These estimates are best made locally.

Printed in UK for HMSO by Eyre & Spottiswoode Limited at Grosvenor Press, Portsmouth
Dd 239244 C25 12/86